The Rightful Place of Science:
Infrastructure in the Anthropocene

The Rightful Place of Science:

Infrastructure in the Anthropocene

Mikhail Chester
Braden Allenby

Consortium for Science, Policy & Outcomes
Tempe, AZ, and Washington, DC

THE RIGHTFUL PLACE OF SCIENCE:
Infrastructure in the Anthropocene

Printed in Charleston, South Carolina.

The Rightful Place of Science series explores the complex interactions among science, technology, politics, and the human condition.

For information on The Rightful Place of Science series, write to: Consortium for Science, Policy & Outcomes
PO Box 875603, Tempe, AZ 85287-5603
Or visit: http://www.cspo.org

Recent volumes in this series:

Sarewitz, D., ed. 2018. *The Rightful Place of Science: New Tools for Science Policy*, vol. I. Tempe, AZ: Consortium for Science, Policy & Outcomes.

Miller, C. A., and Muñoz-Erickson, T. A. 2018. *The Rightful Place of Science: Designing Knowledge*. Tempe, AZ: Consortium for Science, Policy & Outcomes.

Model citation for this volume:

Chester, M., and Allenby, B. 2021. *The Rightful Place of Science: Infrstructure in the Anthropocene*. Tempe, AZ: Consortium for Science, Policy & Outcomes.

ISBN: 0999587781

ISBN-13: 978-0999587782

FIRST EDITION, MARCH 2021

To my father, Dr. Mitchell Chester, who imparted upon me the value of perspective change that comes from curiosity, and a commitment to future generations. My dad used to ask me after school whether I had any homework and if I answered no he'd suggest I assign myself some. This book is one of those assignments.

Mikhail Chester

To my mother and father, who made me the adult I try to be, and to my wife and children, who help me be the imaginative child I need to be.

Braden Allenby

CONTENTS

1

WELCOME TO THE ANTHROPOCENE

Braden Allenby

Infrastructure appears to be caught between the past and the future, and when it comes to these obdurate systems, the present is not a good place to be. Technological evolution is accelerating, politics and society are fragmenting, climate and the natural world are changing, and infrastructure (at least in the developed world) is in dire need of modernization. As countries around the world deploy new infrastructure, the initiatives largely perpetuate the physical forms that we've seen for decades and reflect the same financing structures, management structures, and relationships with the surrounding natural environment as in the past.

Today's design principles seem to reflect those of the last century, when the types of services we demanded of our infrastructure were stable—just like the climate and technological conditions under which infrastructure had to be reliable. But stability is no longer the norm. Furthermore, many forces critical to infrastructure design and performance (such as climatic conditions, advanced technologies such as artificial intelligence, and, less obviously, security protections from today's conflicts) appear to be changing in unpredictable ways. The inflexible and long-lasting nature of infrastructure is directly at odds with these dynamic forces.

A Note on Terminology

We think of infrastructure broadly as consisting of human design spaces. However, the focus of this book is on a particularly important set of engineered infrastructure systems that are designed, built, and managed to deliver basic and critical services, such as energy, water, mobility, and information. These sociotechnical systems are composed of physical assets and the institutions that manage, govern, finance, and regulate them, as well as the massive educational enterprise that creates, stores, and disseminates knowledge to infrastructure engineers and managers. The services these infrastructures provide are not purely physical. We acknowledge that the term "infrastructure" has many meanings and can refer to physical or non-physical (e.g., governance) systems. Our perspective is often on U.S. or modern infrastructure, but we acknowledge that regions of the world are in various stages of infrastructure development. Regardless of where a region is in deploying infrastructure, we believe the principles presented in this book are valuable.

Evidence has been accumulating for decades that we are at the dawn of major shifts in the relationships among humans, social systems, the environment, and technology. This has profound implications for how we design and manage infrastructure, which for a long time policymakers and the public have been mostly able to ignore. But no more.

Hyperconnectivity and the embedding of new technologies into legacy systems, artificial intelligence managing how we understand and demand services from infrastructure, and destabilizing climate extremes represent just a few of the emerging realities that infrastructure managers will face. The rate at which we are deploying new technologies embedded within infrastructure appears to be outpacing the infrastructure itself. A single roadway intersection may have experienced a progression of control systems that started with a traffic officer, progressed to loop detectors, became traffic cameras with image recognition, and is now driven by cloud-based systems such as Google Maps.

Several trends are emerging that affect how we think about and manage infrastructure into the future. First, while change has occurred across social, environmental, and technological systems over the history of humankind, the acceleration of change appears ready to take off: technological development in the coming century is expected to be greater than the past several thousand years. This has profound implications for the planet and human-managed systems.

Second, "nonstationarity," meaning that past trends are no longer good predictors of future conditions, has emerged as an important concept in climate science. This concept can also be applied across a number of systems that affect infrastructure.

Third, infrastructures and the systems that support them have in many ways become so complex that their emergent characteristics—i.e., what we expect them to do when perturbed—are no longer predictable.

Fourth, infrastructure has become a battleground between adversaries in new forms of war that pose huge challenges but are not yet perceived, much less understood, by either the public or infrastructure managers. Yet the training that educational institutions deliver for future infrastructure managers focuses mainly on elements of the system and their predictability (tools for a complicated, not complex, system).

These forces are combining to create conditions of unpredictability that are inimical to the approaches that we've used to design and manage infrastructure in the past. As these forces emerge concurrently, they represent an environment of destabilization that requires new thinking and competencies for how we approach infrastructure into the future. We summarize these trends in Table 1.

Table 1. Trends Affecting Infrastructure Across Domains

DOMAIN	*Acceleration*	*Nonstationarity*	*Emergence*
INFRASTRUCTURE	Quickening integration of cyber technologies into legacy infrastructure, and introduction of information sense-making organizations (Uber, Lyft, Tesla, Google Nest, etc.) that distribute control and affect service consumption.	Design norms focused on stability increasingly insufficient with unpredictable climate-related extreme events.	Increasing complexity due to layering of technologies (accretion, interactions, edge cases, and common rarities).[1]
CYBER-TECHNOLOGIES	Acceleration of technologies where cycle times exceed that of infrastructure. Decreasing of technology costs, increasing data processing capabilities, increasing of communication capabilities.[2]	Cybersecurity in an age of civilizational conflict, where infrastructure is a primary target.[3]	Artificial intelligence; social media tribe formation; China's social credit system.

BIOTECH	Integration of AI/big data/analytics, development of accurate gene editing techniques, and increasing importance of biological data for national security purposes and state AI capabilities significantly accelerate state of art.	As humans become a design space, they are both the designers and the designed, creating a rapid reflexivity that affects all human systems dramatically and unpredictably.	The human in all domains – physical, psychological, identity, community – becomes a design space.
POLITICS	Integration of information and communication technology and AI/big data/analytics with political activities dramatically increases speed of political change.	Increasing ideological polarization stalls infrastructure policy.[4] Increasing contingency of world order and Western Universalism morality significant source of new complexity. Rise of non-traditional political entities leads to growth of neomedieval governance structures.[5]	End of traditional Enlightenment politics leading to unpredictable shifts in geopolitical power, possible ascendency of soft authoritarianism over classic Western pluralism.
FINANCING	New players investing in and building their own systems to circumvent the legacy systems; e.g., Amazon and drone delivery.	Increasing needs for infrastructure financing and restructuring of financing, yet large uncertainty about consistent infrastructure investment.	Increasing tying up of infrastructure financing with other goals.
EARTH SYSTEMS, INCLUDING CLIMATE	Accelerating effects of human activity on water, nutrient, resource, and climate systems, both unintentionally and intentionally.	Nonstationarity in environmental conditions (water, temperature, fire, etc.) that threatens the basic design assumptions of infrastructures and their reliability.[6]	Feedback loops that future climate change will reduce the efficiency of the Earth system to absorb anthropogenic carbon.[7] Emergence of integrated systems that are not well managed under current governance regimes.

SOCIAL	Changing demographics. Software increasingly steering decision-making. Rise of China and Confucianism challenge post-WWII Western universalist ethics and institutions.	Success of Enlightenment has created complexity that in turn undermines Enlightenment concepts, institutions, and governance mechanisms.	Civilizational conflict results in tribal narratives and individual identity becoming both design spaces and battlespaces. Weaponized narrative and manipulation of individuals evolves into new normal for politics.
CULTURAL	Changing values, customs, and beliefs. Changing patterns of war and conflict.[8]	Cultural shifts as demographics change (the old get older, and the poor get poorer).	Cultural homogeneity at globalist level replaced by regional cultural hegemons as new social structures evolve, but destabilize existing social and cultural practices.

For infrastructure designers, managers, and even users, it's time to rethink the relationship with the core systems that serve as the backbone for virtually every activity and service that society demands. It's time to come to grips with the reality that the complexity of infrastructure is exploding, that emerging and disruptive technologies are accelerating in ways that are antithetical to current infrastructure's core design principles, and that education on these issues is insufficient. New accelerating and interactive forces are defining what infrastructure can and should do, and how it should function on a planet dominated by human systems. To understand these accelerating forces and complexity, it's helpful to start in the past.

A Brief History of Infrastructure

Humans have been changing the Earth for millennia. The development of agriculture some ten thousand years ago reflected drastic technological, cultural, economic, and social change, and over time supported a significant jump in global human population from perhaps a million to hundreds of millions of people. It also affected natural systems, from water and land use patterns to biodiversity, to patterns of nutrient and material flows, including those of many metals.[9] It began a process of urbanization that has continued to characterize the human species—and change the planet in numerous ways.[10]

But it was the Industrial Revolution in Europe in the 1700s, and concomitant changes in integrated and co-evolving human, natural, and built systems, that marked the real emergence of the Anthropogenic Earth.[11] The shift from hunter-gatherer to agricultural lifestyles transformed local and regional systems and generated new patterns of resource consumption and energy and waste flows; but the magnitude of those impacts increased exponentially with the Industrial Revolution.

The result over a period of a few short centuries was the terraforming of the planet. Not only did human systems—from economics to technology to culture—change in unpredictable and fundamental ways, but the dynamics of virtually all major natural systems were increasingly affected by human activity.[12] The Industrial Revolution fostered a rapid acceleration in the growth of energy and water use, environmental impacts of all kinds, human population levels and urbanization, economic growth, technological complexity, and the built infrastructure to support it all. These patterns show little signs of slowing.[13]

This acceleration includes not only global population growth, from roughly 450 million people in 1500 to over 7.5 billion today, but also economic growth. Between 1500 and 1800, the global economy grew by a factor of almost three, but between 1500 and today, it has grown by a factor of 12, and much of that growth has been uneven. Per-capita gross domestic product (GDP) grew from 1500 to today by a factor of 10.[14] And with these population and

GDP explosions came exponential growth in technologies, resource use, infrastructure, and environmental impacts.[15]

Such growth rates, which are continuing, imply dramatically expanded interactions between human and natural systems, frequently based on fundamentally new technology systems such as railroads, electricity, and the internet. They also explain the shift from a planet where humans are but one species among many to a world increasingly shaped by the activities of, and for the purposes of, a single species.[16] In such a world, natural cycles and systems transmute from exogenous conditions into infrastructure components.

The growth experienced during this Industrial Era can be framed from a complex adaptive systems perspective. Economies grew and reorganized naturally, and infrastructure grew to support those activities. Energy is sometimes used as the core resource to describe the economic growth of complex adaptive systems. Early humans had few technologies to harness the abundance of fossil and renewable energy around them. The development of technologies and specializations that grew rapidly during the agricultural era and exploded during the industrial era to access and better use energy resulted in increasing complexity in human societies. Agricultural societies relied largely on free solar energy; in the Industrial Era society transitioned to fossil fuels.

Ultimately, as the cost of local energy escalates, economic activity switches to less energy-intensive services and relies on manufacturing and resources from other countries.[17] Throughout these transitions, infrastructures are designed and deployed to provide services that are driven by these activities, many of which rely on the direct consumption of resources (energy, water, materials, etc.). Where abundant resources exist, infrastructure is often "grown" to consume them. Indeed, the Industrial Revolution and subsequent global economic and human population growth reflects the greater availability of energy, which is both an infrastructure and a resource issue. Thus, for example, in the United Kingdom water power was the original energy source in the early Industrial Revolution, but it was rapidly augmented, and then replaced, by fossil fuel use—first coal and then petroleum.[18]

Looking at the processes of growth of complex adaptive systems through the lens of resource exploitation and the managing of unintended consequences has led some scholars to hypothesize about the inevitability of collapse.[19] Nonetheless, although the environmental and energy implications of industrialization and global development have encouraged such dystopian scenarios, economic and demographic growth is a mixture of human institutions and cultural factors; environmental and resource issues and constraints; and interdependent, coevolving technological and economic factors.[20] It is not clear what "collapse" means in the context of such evolving systems.

For example, it is now accepted that the "fall of Rome" was not the collapse of Western civilization at the hands of barbarian tribes, but a shift in form that enabled the subsequent rise of modern civilization.[21] Infrastructure systems are complex precisely because single-domain determinisms and ideological certainties fail; in the end, it is not the belief system, but whether the infrastructure works, that marks an engineering and institutional success. And, especially in the future, the ability to integrate across human, natural, and built systems, with all the concomitant complexity, requires a fundamental shift in how we view the relationship between human and environmental systems.

The rapidly increasing pace of technological change, human culture, and built environments, coupled with their interactions with natural systems, has produced novel and highly complex emergent behaviors that require us to think differently about infrastructure.[22] This complexity is the result of rapidly coevolving human and natural systems.[23] The changes in technological, human, built environment, and natural systems, and the complex outcomes they produce are not intentional. Changes in climate, biodiversity, nutrient cycles, and microbial evolution are unplanned dynamics, the result of technology, policy, and cultural shifts that have been accumulating for generations. But given the emergence and global scale of environmental challenges, the notion that infrastructure should be limited to local engineered systems must be challenged.

Reflecting this shift to a planet dominated by human impacts and activities, scientists have proposed the term "Anthropocene," from the Greek for "human" ("Anthropo-") and "new" ("-cene"), as an appropriate name for the current geologic period. This is not a new idea: while the term was popularized in an article in 2000 entitled "The 'Anthropocene,'"[24] the concept underlies earlier texts such as *Man's Role in Changing the Face of the Earth* (1956)[25] and *The Earth as Transformed by Human Action* (1993).[26] Indeed, as early as 1873 the Italian priest, geologist, and paleontologist Antonio Stoppani used the term "anthropozoic era."[27]

The challenges and opportunities of the Anthropocene require the development of new institutions and frameworks that are beyond today's traditional disciplinary structures and reductionist approaches.[28] Our current infrastructure institutions compartmentalize knowledge and emphasize disciplinary expertise with a focus on components.[29] In contrast, the anthropogenic Earth will be characterized by rapidly evolving technologies, human and natural systems, and information access. This will require us to develop new capabilities to analyze, design, engineer, construct, maintain, manage, and reconstruct infrastructure.[30] The increasing complexity and rates of change require new approaches and sophistication for infrastructure.[31] Sustainable engineering and associated domains;[32] industrial ecology and associated methodologies such as life-cycle assessment (LCA) and materials-flow analysis;[33] systems engineering;[34] adaptive management;[35] as well as relevant parts of the urban planning and sociology of technology[36] are each positioned to support this complexity.

The institutions and engineering disciplines that today are responsible for infrastructure are still appropriate for many challenges. In this book we use the term "infrastructure" to encompass a plurality of parts (both physical and institutional) that provide services that enable human capabilities.[37] The discussion largely focuses on infrastructures as physical systems. The competencies for designing and building a bridge, jet turbine, or microchip are successfully taught and practiced across engineering. However, when it comes to complex and integrated systems, our infrastructure institutions and training are ill-prepared.[38]

At the beginning of the Anthropocene our systems are becoming increasingly integrated, complex, and defined by a greater number of stakeholders with competing priorities. We find ourselves with little training or capability to perceive or parse such systems to design, build, operate, refurbish, or retire infrastructure. This inability is tied to complex governance, fiscal, and educational systems (to name a few) with substantial inertia over time. Codes, conventions, processes, and procedures get developed and are embedded in legal processes and instruments that take considerable time and effort to change. Moreover, different groups (professional associations, government bureaucrats, engineering firms, and university courses) are invested in particular ways of doing things.[39] The complexity of the effects of engineered systems is becoming apparent, even if we don't have the training to fully understand it.

Transportation infrastructure development, for instance, must consider shared, electric, and connected vehicles; power infrastructure, new battery storage technology and rapid advancement of renewables; and water infrastructure and climate change. Additionally, considerations for social equity, cultural impacts, aging populations, terrorism, and the increasing connectedness of hardware and embedding of software must be included.

As natural systems increasingly become design spaces for infrastructure, new approaches are needed for planning, constructing, operating, rehabilitating, and decommissioning infrastructure. As the scale, scope, and technologies of human activities have accelerated, the reductionist approach when assessing the relationship between infrastructure and the environment is no longer acceptable.[40]

We posit that fundamental challenges to managing infrastructure exist in both education and design,[41] arguing that in the short term training should include competencies in complex systems, big data, artificial intelligence, and cybersecurity. In the medium term new approaches are needed that emphasize agility and flexibility.[42] And in the long term we must embrace the complexity inherent in the infrastructure-environment interface and evolve our infrastructure institutions to embrace this complexity.

It is not our intention to exhaustively explore the ways in which infrastructure design and management should change, but instead to establish the need to rethink the relationship between infrastructure and the natural world, and the fundamental challenges for meeting human needs in the coming century. In doing so, we do not take the position that *understanding* leads to *control*. On the contrary, with complex adaptive systems, it is the inability to control the system that is the problem[43] – one that requires the embrace of uncertain emergent behaviors.

Principles for Infrastructure in the Anthropocene

In this book we organize several existing essays into a vision of the challenges and associated restructuring that is needed to design and operate infrastructure in the Anthropocene. The essays were written between 2018 and 2019 and published largely as journal publications. We cite these publications in the text. We felt that a structuring of these essays into a coherent vision was valuable—and even necessary—to begin the conversation of how society takes a different approach to infrastructure going forward. It was especially important that this vision be accessible to the next generation of infrastructure managers. As such we sought to keep this book concise and affordable.

We present several principles, challenges, and recommendations for reimagining how we approach infrastructure. In Chapter 2, we argue that the scale and scope of human activities has grown so large that the dichotomy between infrastructure and the environment is shrinking and often nonexistent; reductionist approaches that approach infrastructure as hardware problems are insufficient. In Chapter 3, we tackle the question of how we should be reimagining infrastructure in the Anthropocene as a wicked and complex process that is less about hardware and more about the management of problems that have no clear solution and are steeped in uncertainty. In Chapter 4, we explore why infrastructure systems remain so rigid, which works against our capabilities to meet demands in rapidly changing, uncertain, and complex environments, and what competencies are needed to make them agile and flexible.

Next, in Chapter 5, we discuss these implications in the context of climate change, and how uncertainty and volatility in climatic conditions are problematic for infrastructure designed for stability. Chapter 6 explores how the scale and scope of human activities have grown so fast that the dichotomy between infrastructure and the environment is disappearing. In Chapter 7, we explore the accelerating integration of cybertechnologies and information into legacy infrastructure systems—resulting in new capabilities but also vulnerabilities. Lastly, in Chapter 8, we argue that infrastructure systems should be approached as wicked complex processes and not simply amalgamations of hardware, and that the capabilities and types of infrastructure will need to differ from those emphasized today.

Notes

[1] S. Arbesman, *Overcomplicated: Technology at the Limits of Comprehension* (New York, NY: Penguin Publishing Group, 2016); M. V. Chester and B. Allenby, "Infrastructure as a wicked complex process," *Elementa* 7, no. 1 (2019): 21.

[2] R. Kurzweil, *The Singularity Is Near* (New York, NY: Viking, 2005); R. Rajkumar, "A Cyber–Physical Future," *Proceedings of the IEEE* 100 (2012): 1309–1312; B. Allenby, "5G, AI, and big data: We're building a new cognitive infrastructure and don't even know it," *Bulletin of the Atomic Scientists* (Dec. 19, 2019).

[3] B. R. Allenby, "The paradox of dominance: The age of civilizational conflict," *Bulletin of the Atomic Scientists* 71, no. 2 (2015): 60–74.

[4] "The Partisan Divide on Political Values Grows Even Wider," Pew Research Institute (Oct. 5, 2017).

[5] H. Kissinger, *World Order* (New York, NY: Penguin Books, 2014); S. McFate, *The New Rules of War: Victory in the Age of Durable Disorder* (New York, NY: William Morrow, 2019).

[6] H. F. Lins, "A Note on Stationarity and Nonstationarity" (June 2012), available at: https://www.wmo.int/pages/prog/hwrp/chy/chy14/documents/ms/Stationarity_and_Nonstationarity.pdf

[7] P. Friedlingstein, P. Cox, R. Betts, L. Bopp, W. von Bloh, V. Brovkin, P. Cadule, S. Doney, M. Eby, I. Fung, G. Bala, J. John, C. Jones, F. Joos, T. Kato, M. Kawamiya, W. Knorr, K. Lindsay, H. D. Matthews, T. Raddatz, P. Rayner, C. Reick, E. Roeckner, K.-G. Schnitzler, R. Schnur, K. Strassmann, A. J. Weaver, C. Yoshikawa, and N. Zeng, "Climate–Carbon Cycle Feedback Analysis: Results from the C4MIP Model Intercomparison," *Journal of Climate* 19, no. 14 (2006): 3337–3353.

[8] S. McFate, *The New Rules of War: Victory in the Age of Durable Disorder* (New York, NY: William Morrow, 2019).

[9] W. L. Thomas, ed., *Man's Role in Changing the Face of the Earth* (Chicago, IL: University of Chicago Press, 1956); B. L. Turner, *The Earth as Transformed by Human Action: Global and Regional Changes in the Biosphere Over the Past 300 Years* (Cambridge, UK: Cambridge University Press, 1993).

[10] P. Hall, *Cities in Civilization* (New York, NY: Pantheon Books 1998).

[11] J. Syvitski, "Anthropocene: An Epoch of Our Making," *Global Change Magazine* (March 19, 2012): 12.

[12] B. L. Turner, *The Earth as Transformed by Human Action: Global and Regional Changes in the Biosphere Over the Past 300 Years* (Cambridge, UK: Cambridge University Press, 1993); B. Allenby, "Earth Systems Engineering and Management: A Manifesto," *Environmental Science & Technology* 1 (2007): 7960–7965.

[13] B. Allenby, *The Theory and Practice of Sustainable Engineering* (Upper Saddle River, NJ: Pearson Prentice Hall, 2012).

[14] Ibid.

[15] J. Syvitski, "Anthropocene: An Epoch of Our Making," *Global Change Magazine* (March 19, 2012): 12.

[16] B. Allenby and M. Chester, "Reconceptualizing Infrastructure in the Anthropocene," *Issues in Science and Technology* 34, no. 3 (2018): 58–63.

[17] G. Tverberg, "A New Theory of Energy and the Economy - Part 1 - Generating Economic Growth," *Our Finite World* blog, (January 21, 2015).

[18] A. Grübler, *Technology and Global Change* (Cambridge, UK: Cambridge University Press 1998).

[19] J. Tainter, *The Collapse of Complex Societies* (Cambridge, UK: Cambridge University Press, 1990).

[20] N. Rosenberg and L. E. Birdzell, *How the West Grew Rich: The Economic Transformation of the Industrial World*. (New York, NY: Basic Books, 1986); E. D. Beinhocker, *The Origin of Wealth: Evolution, Complexity, and the Radical Remaking of Economics* (Cambridge, MA: Harvard Business School Press, 2006).

[21] J. C. Scott, *Against the Grain* (New Haven, CT: Yale University Press, 2017).

[22] B. Allenby and M. Chester, "Reconceptualizing Infrastructure in the Anthropocene," *Issues in Science and Technology* 34, no. 3 (2018): 58–63.

[23] B. Allenby, *The Theory and Practice of Sustainable Engineering* (Upper Saddle River, NJ: Pearson Prentice Hall, 2012).

[24] P. Crutzen and E. Stoermer, "The Anthropocene," *Global Change Newsletter* 41 (May 2000).

[25] W. L. Thomas, ed., *Man's Role in Changing the Face of the Earth* (Chicago, IL: University of Chicago Press, 1956).

[26] B. L. Turner, *The Earth as Transformed by Human Action: Global and Regional Changes in the Biosphere Over the Past 300 Years* (Cambridge, UK: Cambridge University Press, 1993).

[27] P. Crutzen and E. Stoermer, "The Anthropocene," *Global Change Newsletter* 41 (May 2000); B. Allenby, "Anthropocene," in J. B. Holbrook, ed., *Ethics, Science, Technology, and Engineering: A Global Resource* (Farmington Hills, MI: Gale Cegnage Learning, 2014): 690.

[28] B. Allenby and M. Chester, "Reconceptualizing Infrastructure in the Anthropocene," *Issues in Science and Technology* 34, no. 3 (2018): 58–63.

[29] W. E. Bijker, T. P. Hughes, and T. Pinch, *The Social Construction of Technological Systems: New Directions in the Sociology and History of Technology* (Cambridge, MA: MIT Press, 1987); A. Carse and J. A. Lewis, "Toward a political ecology of

infrastructure standards: Or, how to think about ships, waterways, sediment, and communities together," *Environment and Planning A: Economy and Space* 49, no. 1 (2017): 9–28.

[30] B. Allenby, *The Theory and Practice of Sustainable Engineering* (Upper Saddle River, NJ: Pearson Prentice Hall, 2012).

[31] B. Allenby and M. Chester, "Reconceptualizing Infrastructure in the Anthropocene," *Issues in Science and Technology* 34, no. 3 (2018): 58–63.

[32] O. L. De Weck, D. Roos, and C. L. Magee, *Engineering Systems: Meeting Human Needs in a Complex Technological World* (Cambridge, MA: MIT Press, 2011); B. Allenby, *The Theory and Practice of Sustainable Engineering* (Upper Saddle River, NJ: Pearson Prentice Hall, 2012).

[33] T. E. Graedel, and B. R. Allenby, *Industrial Ecology and Sustainable Engineering* (Upper Saddle River, NJ: Pearson, 2009).

[34] T. P. Hughes, *Rescuing Prometheus* (New York, NY: Vintage Books, 2000); A. Kossiakoff, W. N. Sweet, S. J. Seymour, and S. M. Biemer, *Systems Engineering Principles and Practice* (Hoboken, NJ: John Wiley & Sons, 2011).

[35] C. Folke, J. Colding, and F. Berkes, *Linking Social and Ecological Systems: Management Practices and Social Mechanisms for Building Resilience* (Cambridge, UK: Cambridge University Press, 2000); L. H. Gunderson, C. S. Holling, and S. S. Light, *Barriers and Bridges to the Renewal of Regional Ecosystems* (New York, NY: Columbia University Press 2010).

[36] W. E. Bijker, T. P. Hughes, and T. Pinch, *The Social Construction of Technological Systems: New Directions in the Sociology and History of Technology* (Cambridge, MA: MIT Press, 1987).

[37] A. Carse, "Keyword: infrastructure: How a humble French engineering term shaped the modern world," in P. Harvey, C. Bruun Jensen, and A. Morita, eds., *Infrastructures and Social Complexity* (Oxfordshire, UK: Routledge, 2017), 45–57; S. S. Clark, T. P. Seager, and M. V. Chester, "A capabilities approach to the prioritization of critical infrastructure," *Environment Systems and Decisions* 38 (2018): 339–352.

[38] B. Allenby and M. Chester, "Reconceptualizing Infrastructure in the Anthropocene," *Issues in Science and Technology* 34, no. 3 (2018): 58–63.

[39] R. Bendix, *Max Weber: An Intellectual Portrait* (Berkeley, CA: University of California Press, 1978).

[40] B. Allenby and M. Chester, "Reconceptualizing Infrastructure in the Anthropocene," *Issues in Science and Technology* 34, no. 3 (2018): 58–63.

[41] Ibid.

[42] M. Chester and B. Allenby, "Toward Adaptive Infrastructure: Flexibility and Agility in a Non-Stationarity Age," *Sustainable and Resilient Infrastructure* 3, (2018): 1–15.

[43] B. Allenby, *The Theory and Practice of Sustainable Engineering* (Upper Saddle River, NJ: Pearson Prentice Hall, 2012).

2

EARTH SYSTEMS ENGINEERING AND MANAGEMENT: A MANIFESTO*

Braden Allenby

The Industrial Revolution led to changes in human demographics, agricultural and technology systems, cultures, and economic systems. A principal result has been the evolution of an anthropogenic Earth in which the dynamics of major natural systems are increasingly affected by human activity. That does not mean deliberately designed by humans, because many things, from urban systems to the internet, are clearly human in origin yet have not been consciously designed by anyone. But it does mean an Earth where human activity increasingly modulates all Earth systems to the point where those things that are not subject to such impact, such as perhaps volcanoes and earthquakes, are increasingly limited and rare.[1]

It is a world characterized by rapidly increasing integration of human culture, built environments, and natural systems to produce novel and complex emergent behaviors

---

* Reprinted from B. Allenby, "Earth Systems Engineering and Management: A Manifesto," *Environmental Science and Technology* 41, no. 23 (2007): 7960–7965, with permission from ACS Publications.

that are beyond traditional disciplinary structures and reductionist approaches. As the journal *Nature* put it in a 2003 editorial, "Welcome to the Anthropocene," which roughly translates to "the Age of the Human."[2]

The boundaries reflected in today's engineering disciplinary structures, and indeed in academic systems as a whole, are still appropriate for many problems. But we fail at the level of the complex, integrated systems and behaviors that characterize the anthropogenic Earth. No disciplinary field in either the physical or social sciences addresses these emergent behaviors, and very few even provide an adequate intellectual basis for parsing such complex adaptive systems. This situation has two important implications for civil and environmental engineering (CEE) professionals.

First, it means that we as engineers cannot continue to rest on our traditional strengths, which are increasingly inadequate given today's social, economic, environmental, and technological demands. For example, a road built into a rain forest to support mineral exploitation becomes a corridor of development and environmental degradation. Similarly, a new airport in a developing country dramatically increases tourism and puts pressure on fragile, previously remote, ecosystems. Alternatively, planning for urban transportation infrastructure increasingly requires understanding the status of the information and communication technology (ICT) infrastructure, because ICT enables virtual work structures that affect potential traffic loading and peak patterns. In every case, traditional CEE approaches, although necessary, do not address the systemic impacts of the project. Infrastructure is critical but not neutral.

Second, from a proactive viewpoint, the anthropogenic Earth is a highly complex and tightly integrated system that challenges society to rapidly develop tools, methods, and knowledge that enable reasoned responses. Engineers in general, and civil and environmental engineers in particular, must be a critical part of any such response. As problem solvers who create solutions in the real world, we have to understand and appropriately consider this new and more complex environment within which we work and create future options for changing ecosystems, built environments, and human culture. The rational and analytical CEE culture, along with the role of CEE professionals in creating and maintaining the built environment, makes the CEE community a necessary partner—indeed, leader—in Earth systems engineering and management (ESEM).

Earth Systems Engineering and Management

Continued stability of the information-dense, highly integrated human, natural, and built systems that characterize the anthropogenic Earth requires the ability to rationally design, engineer and construct, maintain and manage, and reconstruct such systems—in short, an ESEM capability.[3] Although this is an unprecedented challenge, ESEM can draw on experience from many existing areas of study and practice. From a technical perspective, these include industrial ecology methodologies such as life-cycle assessment, design for environment, materials flow analysis,[4] and systems engineering.[5] From a managerial perspective, it draws on the literature about learning organizations[6] and adaptive management.[7] Parts of the urban planning, sociology of technology, and social construction literatures are also relevant.[8]

On the basis of these discourses, a tentative and partial, albeit instructive, set of initial ESEM principles can be developed.

Given our current level of ignorance, *only intervene when necessary, and then only to the extent required, in complex systems.* This follows from the obvious need to treat complex adaptive systems with respect, because their future paths and reactions to inputs can seldom be predicted. It supersedes formulations such as the precautionary principle, which, in holding that new technologies should not be introduced if the risks cannot be known, demands an unrealistic level of knowledge of the future. Moreover, engineers who solve problems in the real world must accept the world as it is—globalizing, growing rapidly economically, with a population of more than 7 billion people, and heavily reliant on technological systems. Intervention is thus not discretionary, as some would rather fancifully wish, but it nonetheless must be careful and rational.

The capability to model and dialogue with major shifts in technological systems should be developed before, rather than after, policies and initiatives encouraging such shifts. Although projections of technological evolution are seldom accurate, we could do much better in developing frameworks, tracking systems (including metrics, especially ones that signal potential danger), and families of scenarios that would help us perceive problematic trends in real time, and perhaps steer technological evolution to increase social and environmental benefits. Such systematic tracking capabilities can help avoid some of the costs of premature adoption of emotionally appealing technologies. Recent examples might include the current infatuation with the hydrogen economy or the massive effort by the United States to create a corn-based ethanol energy economy. The point is not, of course, that technology shifts may not be beneficial; the point is to improve their design and management as they evolve within the real world.

A characteristic of complex systems is that the network that is relevant to a particular analysis is called forth by that analysis. Accordingly, *it is critical to be aware of the particular boundaries within which one is working and to be alert to the possibility of logical failure when one's analysis goes beyond the boundaries.* For example, to perform a study of New York City's water supply by considering only the five constituent boroughs of New York would result in a flawed assessment, because the system being analyzed (water provision to the city) is not mapped adequately by the political boundaries of the city. Similarly, the application of a life-cycle assessment tool that relied heavily on energy consumption as a proxy for environmental damage to a product where toxicity was a primary issue might well result in dysfunctional conclusions. For example, replacing chlorofluorocarbon-cleaning technologies with aqueous ones in electronics manufacturing makes sense from a systems perspective, even though the latter is more energy-intensive.

A point that is critical to an understanding of the anthropogenic world is that *the actors and designers are also part of the system they are purporting to design, creating interactive flows of information (reflexivity) that make the system highly unpredictable and perhaps more unstable.* As scientists develop data on the effects of global climate change, for example, people's perceptions are changed. This, in turn, changes social practices affecting the climate. Thus, activities at the levels of the emergent behaviors of these complex systems must be understood as processes and dialogues, rather than simply problems to be solved and forgotten. This is an issue that bifurcates engineering: most engineering still involves artifacts, but ESEM requires ongoing and highly sophisticated dialogues with the systems at issue.

Implicit social engineering agendas and reflexivity make ethical and value implications inherent in all ESEM activities. To achieve long-term clarity and stable, effective policies, these normative elements must be explained and accepted, rather than hidden.

Conditions characterizing the anthropogenic Earth require democratic, transparent, and accountable governance and pluralistic decision-making processes. Virtually all ESEM initiatives raise important scientific, technical, economic, political, ethical, theological, and cultural issues in an increasingly complex global polity. Given the need for consensus and long-term commitment, the only workable governance model is one that is democratic, transparent, and accountable.[9]

We must learn to engineer and manage complex systems, not just artifacts. An obvious result of the above analysis is that the anthropogenic world—and ESEM as a response—requires that far more attention be paid to the characteristics and dynamics of the relevant systems, rather than just to constituent artifacts. This does not negate the need to design artifacts; ESEM augments, instead of replaces, more traditional activities.

Ensure continuous learning. Given the complexity of the systems involved, our relative ignorance, and the recognition of engineering as process, it follows that continual learning at the personal and institutional level must be built into project and program management. Some experience with this approach already exists. High-reliability organizations, such as aircraft carrier operations or well-run nuclear power plants, usually have explicit learning structures.[10] Similarly, the adaptive management approach to complex natural resource management challenges, such as in the Baltic Sea, the Everglades, and the North American Great Lakes, is heavily dependent on continual learning.[11]

Unlike simple systems, complex, adaptive systems cannot be centrally or explicitly controlled. Accordingly, it's important to understand not just the substance of the system—the biology of the Everglades or the Baltic, for example, or the physics and chemistry of the troposphere—but also inherent systems dynamics. Where in a system do small shifts propagate across the system as a whole, and where are they dampened out? The famous example of the butterfly that flaps its wings and causes a storm elsewhere in the world may be iconic, but what is perhaps forgotten is that millions of butterflies flap their wings thousands of times each day, without causing an ensuing storm. Perhaps the really interesting question, then, is why one flap has such an impact, when the others don't.[12]

Whenever possible, engineered changes should be incremental and reversible, rather than fundamental and irreversible. Accordingly, premature lock-in of system components should be avoided where possible, because it leads to irreversibility. In complex systems, practices and technologies can get locked in quickly—that is, coupled to other systems and components in such a way as to make subsequent changes, including reversion to previous states, difficult or impossible. Thus, tightly coupled networks are more resistant to change than loosely coupled networks, an effect that can be reduced by ensuring that, when couplings to other networks do exist, they are designed to be as loose, and as few, as possible. This supports the more general goal of reversibility: under conditions of high uncertainty and complexity, easy reversibility is a desirable option should the system begin to behave in an unanticipated and undesired way.

ESEM projects should aim for resiliency, not just redundancy, in design. Redundancy provides backup capability in case a primary system fails, and it is commonly designed into high-reliability systems such as jet airplanes. Redundancy assumes, however, that the challenge to the system is of a known variety. Resiliency, to the contrary, is the ability of a system to resist degradation or, when it must degrade, to do so safely even under unanticipated conditions.[13]

Developing an ESEM Capability

One way to begin responding to the challenge of the anthropogenic Earth, as well as continuing the process of clarifying and better understanding ESEM, is to develop a model research agenda. The complexity of the challenges does not allow for more than a partial and exploratory exercise at this point, and the examples given below are also idiosyncratic in that they reflect a CEE perspective on ESEM. In addition, it is a legitimate concern that any discipline, including CEE, that attempts to train professionals to design, engineer, manage, and interact with such complex systems is doomed to overreach and fail. Nonetheless, it is also important to remember that these effects, from climate change to massive urbanization, are already occurring. Our failure to accept responsibility for them does not diminish human impacts but is merely an evasion of our ethical duties. CEE has an important role here: its projects are frequently the vehicle by which these large and complex systems are affected, and CEE education – rational, quantitative, problem-oriented, systems-based, and pragmatic – is a solid base upon which to build the required expertise and insight.

Accordingly, in addition to its specific research goals, any ESEM research agenda should aim to support the development of highly transdisciplinary research programs capable of looking at Earth systems at emergent levels (including, importantly, the social science dimensions; ideology and politics are often as important as any physical feature of the system). It should also support an overarching program that mines specific research areas for general principles and learning that over time can be leveraged into development of a rational, responsible, and ethical ESEM framework.

Integrated Urban Infrastructure Systems

Given accelerating urbanization,[14] increasing urban vulnerability to natural disaster or deliberate attack, and the complexity of urban systems, the emergent domain of urban infrastructure systems as comprehensive wholes is grossly underappreciated. Yet, at this point no U.S. government agency, research funding organization, or engineering discipline has the mission or research support for understanding urban systems as integrated systems.

This is a near-term concern because of the increasing demand for new and replacement infrastructure. At the same time, the nature of urban systems is changing profoundly as ICT capability is increasingly integrated into all levels of urban functionality: sensor systems, smart materials, smart buildings, smart infrastructures, and the like. Especially as ICT systems are redesigned to be autonomic—virtualized, self-defining, self-monitoring, self-healing, and learning-capable at all scales from chip to computer to regional and global grids[15]—the implications for urban system design, performance, and behavior accelerate in complexity. Moreover, the increasing role of urban systems as nodes in energy, financial, and virtual information networks adds many layers of information complexity to urban environments.[16] Such research should contribute to a new CEE competency in urban-scale systems design and management.

Sustainable Infrastructures

Growing populations, economic development, accelerating technological change, urbanization, and aging and failing infrastructure systems are increasing the need for sustainable infrastructure systems. Although ESEM provides some conceptual basis for developing such systems, clearly the translation of social interest in sustainability to the implementation of sustainable engineering of any type has just begun. It's currently marked by an intellectually confused jumble of superficial, ideological, and heuristic approaches.[17]

Accordingly, a research program to help define sustainable infrastructure and to develop appropriate methodologies, analytical methods, and tools is needed. This is urgent because the time to understand and deploy sustainable infrastructures is now, instead of after newly built environments with decades of active life are constructed. The new National Science Foundation initiative called Resilient and Sustainable Infrastructures, under the Emergent Frontiers in Research and Innovation program, is clearly a step in the right direction, but whether it leads rapidly to the large transdisciplinary effort to solve the problems of urban environments remains to be seen.

Technological Convergence

A number of authors, from the dystopian Bill Joy[18] to the techno-optimist Ray Kurzweil,[19] have written about the subject of technological convergence, generally understood as including the accelerating development of the fields of nanotechnology, biotechnology, information and communication technology, applied cognitive science, and robotics as well as their mutually reinforcing integration. These converging technologies constitute major Earth systems in their own right, and their complexity and challenging philosophical, religious, ideological, and economic implications are just beginning to be recognized. However, some of the major arenas where effects of technological convergence can first be seen are in areas familiar to CEE professionals. These include urban and regional integrated infrastructure design, energy technology and infrastructure design, and the like.

What is most challenging, perhaps, about technological convergence is not just its effect of turning natural systems—from the carbon and climate cycles to biology at all scales—into design spaces (and commodities). Rather, as humans gain the tools to design biological and cognitive systems, it also turns the human into a self-reflexive design space. In doing so, the feedback systems, and concomitant increases in system complexity, become truly daunting. CEE traditionally has been based on the assumption (unspoken because it was so clearly fundamental and valid) that the environment must be designed and built for the human. As both parts of that assumption become design spaces, and thus interact in new and dynamic ways, engineering becomes new, more complex, and ethically challenging in ways that have never before been part of our professional experience. Although research programs designed to respond to this unique challenge do not lie entirely within CEE's ambit, we can bring significant skills to transdisciplinary research efforts.

Resilience of Complex Earth Systems

That complex natural Earth systems are vulnerable to destabilization as a result of human activity is evident from the depletion of stratospheric ozone by chlorofluorocarbons or from the global climate change dialogue. But the vulnerability is also apparent with anthropogenic systems: recent years have seen major challenges to social stability and order, ranging from extreme weather events to terrorist attacks to substantial cultural conflict. Although each incident is unique and too often tragic, the key to understanding and responding to these constellations of challenges is to recognize that although each is expressed uniquely, they all represent emergent characteristics of the anthropogenic Earth—including, critically, information and cultural networks—at unfamiliar scales and levels of complexity. Thus, although immediate responses have necessarily relied primarily on specific engineering, institutional, and policy responses to particular incidents, the range of challenges, their systemic nature, and the practical impossibility of adequately addressing each one individually argue for adopting a more comprehensive systems perspective. This should be based on the principles of enhancing infrastructure, social, and economic resiliency; meeting security and emergency response needs; and relying to the highest extent possible on dual-use technologies that offer societal benefits, even if anticipated disasters never occur.

Patterns of the built and human environments play an important role in vulnerability. Thus, for example, the damage and disruption from weather events such as hurricanes or from natural disasters such as tsunamis are more disruptive and extensive than in the past because of changing demographic patterns (urbanization, for example) and the relocation of economic activity near riskier areas, such as geologically active Pacific Ocean coastlines. Disease epidemics and their associated economic and social effects are more challenging given the modern transportation infrastructure and globalized patterns of commerce and travel. Terrorism is not new, but terrorist access to weapons of mass destruction is. Cultural conflict is as old as historical records, but the internet and social media create an environment where a few cartoons in a small northern European country can ignite global unrest.

CEE professionals have important roles in virtually all of these examples, including designing adequate levees; hardening buildings and infrastructure against attack and enabling rapid restoration of services and the built environment; and constructing energy, transportation, and ICT infrastructures that have profound and varied effects across regional and global natural systems. We should thus be leaders in enabling systemic understanding and enhancement of resilience across not just the built environment but also Earth systems as a whole. This is a substantial challenge: how can we, as the CEE community, begin the complex process of building the transdisciplinary capabilities necessary to understand, work, and live rationally, ethically, and responsibly in the world that we have already created?

These research challenges, and many others that undoubtedly come to mind, are "wicked" problems, because they are irreducibly complex and highly transdisciplinary, and require substantial changes in the way we think about CEE and engineering in general. Learning to work across the disciplinary divides involved will be exceedingly difficult and personally challenging for many individuals. Many engineers are not accustomed to accepting a leadership role in such a difficult task, but our age has its own imperative.

Activity in each area will be complicated because not enough trained individuals are available to begin many programs in these areas. Peer review will also be a challenge, both because finding appropriate panels will be nontrivial and because that process tends to be highly conservative when faced with profoundly transdisciplinary proposals. In many cases, ideological and even religious feelings run high.

But against all of these barriers lies one fact: we do not have a choice. These emergent behaviors are here, now. We only can decide as engineers and professionals whether to respond to these behaviors rationally and ethically or to ignore them, retreating to wishful fantasy and evading our professional and, indeed, personal responsibility to ourselves and the future.

Notes

[1] B. R. Allenby, *Reconstructing Earth* (Washington, DC: Island Press, 2005).

[2] "Welcome to the Anthropocene," *Nature* 424 (2003): 709.

[3] B. R. Allenby, "Earth Systems Engineering and Management" *IEEE Technology and Society Magazine* 19, no. 4 (2000): 10–24.

[4] T. E. Graedel and B. R. Allenby, *Industrial Ecology,* 2nd ed. (Upper Saddle River, NJ: Prentice Hall, 2003).

[5] R. Pool, *Beyond Engineering: How Society Shapes Technology* (Oxford, UK: Oxford University Press, 1997); T. P. Hughes, *Rescuing Prometheus* (New York, NY: Pantheon, 1998).

[6] P. M. Senge, *The Fifth Discipline* (New York, NY: Doubleday, 1990).

[7] L. H. Gunderson, C. S. Holling, and S. S. Light, eds., *Barriers and Bridges to the Renewal of Ecosystems and Institutions* (New York, NY: Columbia University Press, 1995); F. Berkes, C. Folke, eds., *Linking Social and Ecological Systems: Management Practices and Social Mechanisms for Building Resilience* (Cambridge, UK: Cambridge University Press, 1998).

[8] P. L. Berger and T. Luckmann, *The Social Construction of Reality: A Treatise in the Sociology of Knowledge* (New York, NY: Anchor, 1967); W. E. Bijker, T. P. Hughes, and T. Pinch, eds., *The Social Construction of Technological Systems* (Cambridge, MA: MIT Press, 1997).

[9] J. Habermas, *Legitimation Crisis*, T. McCarthy, trans. (Boston, MA: Beacon Press, 1975); R. Rorty, *Contingency, Irony, and Solidarity* (Cambridge, UK: Cambridge University Press, 1989).

[10] R. Pool, *Beyond Engineering: How Society Shapes Technology* (Oxford, UK: Oxford University Press, 1997).

[11] L. H. Gunderson, C. S. Holling, and S. S. Light, eds., *Barriers and Bridges to the Renewal of Ecosystems and Institutions* (New York, NY: Columbia University Press, 1995); F. Berkes, C. Folke, eds., *Linking Social and Ecological Systems: Management Practices and Social Mechanisms for Building Resilience* (Cambridge, UK: Cambridge University Press, 1998).

[12] S. A. Kauffman, "What Is Life?" in *The Next Fifty Years*, ed. J. Brockman (New York, NY: Vintage, 2002), 126–144.

[13] B. R. Allenby and J. Fink, "Toward Inherently Secure and Resilient Societies," *Science* 309 (2005): 1034–1036.

[14] National Research Council, *Cities Transformed* (Washington, DC: National Academies Press, 2003).

[15] AT&T, "AT&T Network Continuity Overview" (2005); IBM, "Autonomic Computing: An Architectural Blueprint for Autonomic Computing," 3rd ed. (2005).

[16] M. Castells, *The Rise of the Network Society*, 2nd ed. (Oxford, UK: Blackwell, 2000).

[17] B. R. Allenby, D. Allen, and C. Davidson, "Sustainable Engineering: From Myth to Mechanism," *Environmental Quality Management* 17, no. 1 (2007): 17–26.

[18] B. Joy, "Why the Future Doesn't Need Us," *Wired* 8, no. 4 (2000).

[19] R. Kurzweil, *The Singularity Is Near* (New York, NY: Viking, 2005).

3

RECONCEPTUALIZING INFRASTRUCTURE IN THE ANTHROPOCENE*

Braden Allenby and Mikhail Chester

A fundamental shift is afoot in the relationship between human and natural systems. It requires a new understanding of what we mean by infrastructure, and thus dramatic changes in the ways we educate the people who will build and manage that infrastructure. Similar shifts have occurred in the past, as when humanity transitioned from building based on empirical methods developed from trial-and-error experience, which was sufficient to construct the pyramids and European cathedrals, to the science-based formal engineering design methods and processes necessary for the electric grid and jet aircraft. But just as the empirical methods were inadequate to meet the challenges of nineteenth- and twentieth-century developments, today's

* Reprinted with permission from B. R. Allenby and M. Chester, "Reconceptualizing Infrastructure in the Anthropocene," *Issues in Science and Technology* 34, no. 3 (2018): 58–63.

methods are inadequate to meet the needs of the twenty-first century. Now that we have entered the Anthropocene period in which human activities affect natural systems such as climate, engineers face far more complex design challenges.

The durability of the pyramids and cathedrals is evidence that the inherited wisdom of experience can be adequate for many tasks, but we can also see the disadvantages of working without the formal design tools based on advanced mathematics and engineering science. The choir vault and several buttresses of the Beauvais Cathedral in France collapsed in 1284, only 12 years after its partial completion. Retrospective analysis suggests that the problem was resonance under wind load, the type of stress that could be anticipated only with the formal design methods that were yet to be developed. And when the first iron bridge was built at Coalbrookdale, England, in 1781, the engineers used the same design they would have used for a masonry bridge. Within a hundred years, however, structures such as James Eads's 1867 bridge over the Mississippi River at St. Louis and Gustave Eiffel's 1884 Garabit Viaduct were being formally designed using scientific methods and quantitative design processes that made it possible to predict the performance of the new materials. Personal experience and historical heuristics were replaced by scientific data and quantitative models. Opportunities for infrastructure evolution that would have been impossible without scientific understanding suddenly became available, leading to more efficient, effective, and safer infrastructure, and eventually to entirely new forms of infrastructure such as air travel and electrification.

The engineering tools, methods, and practices necessary for a practicing professional; the engineering education necessary to prepare such a professional; and the way engineering is performed institutionally have all changed fundamentally throughout history as the context within which humans live and prosper has changed. We are now in a period of rapid, unpredictable, and fundamental change that is literally planetary in scope. To date, neither our ideas regarding infrastructure, nor our educational and institutional systems for conducting engineering, have reflected this striking evolution in context. We are well past the point where adaptation is pragmatically, and ethically, required.

It is important to remember that this doesn't mean that all current practice and educational methods are suddenly obsolete. Rather, it means that they are increasingly inadequate, especially when applied to projects that involve complex social, economic, and cultural domains. Designing a water treatment plant will remain well within the capability of today's field of environmental engineering; designing the nitrogen and phosphorus cycles and flows in the Mississippi River basin to reduce the life-depleted "dead zone" in the Gulf of Mexico cannot be done so simply.

The Planet as Infrastructure

That we now live on a terraformed planet is not a new idea. The term "Anthropocene" was popularized in an article in 2000 by Paul Crutzen and Eugene Stoermer, but as early as 1873 the Italian priest, geologist, and paleontologist Antonio Stoppani used the term "anthropozoic era," and others have used similar language throughout the twentieth century. The force driving the birth of a new era is obvious. A planet that supported roughly 450 million humans in 1500 now supports over 7.5 billion, and because of industrial progress and economic growth, each of those humans has a vastly more consequential impact.

In a blink of geologic time, then, humans left a planet where they were but one species among many and built a world increasingly shaped by the activities of, and for the purposes of, a single species—themselves. It is not that the planet has been deliberately designed by humans, but many human-built systems from transportation to communications have resulted in global changes that were not consciously designed by anyone. Moreover, that human activity is clearly a significant contributor to current dynamics of such natural phenomenon as climate change, biodiversity, nitrogen and phosphorous cycles, microbial evolution, and regional ecosystem parks such as the Everglades does not imply deliberate design in such cases either. Rather, the world we find ourselves in is one where deliberate human activity, itself a complex network of functions, interacts with a highly complex planetary substrate to create unpredicted, often challenging, emergent behaviors, of which climate change is but one example.

These emergent properties would not exist but for humans; the design process involved, however, is not the explicitly rational and quantitative method that we are used to. Rather, a planet characterized by rapidly increasing integration of human culture, built environments, and natural systems producing novel, highly complex, rapidly changing emergent behaviors challenges all our existing ideas about design, operation, and management of infrastructure. Indeed, it challenges the very idea of infrastructure as limited to local, highly engineered systems. For in such a world, "natural" cycles and systems transmute from exogenous conditions into infrastructure components, a process implicitly recognized by, for example, the substantial literature on ecosystem services and earth systems engineering and management.

Consider urban water supply. Replacing a water pipe in New York City, albeit expensive and complex, is well within current professional and institutional capabilities. On the other hand, to maintain that water supply means designing, through engineering and land use regulation, a continuously monitored network of 19 reservoirs in a roughly 2,000-square-mile watershed. Similarly, most people know that Arizona's population is concentrated in several large urban areas in semiarid desert regions of the state. Contrary to popular belief, however, Arizona's water supply is robust. It is very diverse, with 17% from in-state rivers and associated reservoirs, 40% from groundwater, 3% reclaimed, and 40% imported from the Colorado River, which channels water from seven states. Flexibility and resilience in operations is enhanced because imported water is largely banked in underground reservoirs, while wastewater is highly managed and either returned to the ground where it can be accessed later or sent to power generation facilities. Similarly, water in California is tightly designed. Mountain snow releases water as it melts, and that runoff is heavily managed for multiple uses as it moves toward the ocean. As in many other regions, natural flows from precipitation are now highly managed and there is increasing recognition that historical flow patterns are becoming less and less relevant for predicting future flows. In short, watersheds are now so highly managed that the environment has in effect become the water infrastructure.

As these examples suggest, the advent of the Anthropocene requires the development of new institutions and frameworks that are beyond today's traditional disciplinary structures and reductionist approaches. Continued stability of the information-dense, highly integrated, rapidly evolving human, natural, and built systems that characterize the anthropogenic Earth requires development of new abilities to rationally analyze, design, engineer and construct, maintain and manage, and reconstruct and evolve such systems at local, regional, and even planetary scales. It is not that current institutions and disciplinary boundaries are completely obsolete; rather, it is that increasing complexity and rates of system evolution require a new and more integrative level of sophistication in infrastructure conceptualization, design, and management. This can draw on a number of relatively new fields of study, including sustainable engineering, industrial ecology and associated methodologies such as life-cycle assessment and materials flow analysis, systems engineering, adaptive management, and parts of the urban planning and sociology of technology literatures.

The boundaries reflected in current engineering institutions, disciplines, and educational practices and structures are still appropriate for many problems. But today, at the dawn of the Anthropocene, they increasingly fail at the level of the complex, integrated systems and behaviors that characterize the anthropogenic Earth. No disciplinary field in either the physical or social sciences addresses these emergent behaviors, and very few even provide an adequate intellectual basis for perceiving, much less parsing, such complex adaptive systems. Engineering cannot continue to rest on traditional curricula and strengths, which are increasingly inadequate given today's social, economic, environmental, and technological demands. A road built into a rain forest to support petroleum extraction operations cannot be well designed unless it is understood that such infrastructure becomes a corridor of development and environmental degradation. Similarly, a new airport planned in Peru in the Inca Sacred Valley cannot be adequately designed unless its role in dramatically increasing tourism and development in a fragile ecosystem and historic cultural region is understood, and the numerous effects evaluated. Planning for urban transportation infrastructure is inadequate unless it includes consideration of the implications of such varied developments as hybrid, electric, and autonomous vehicle technologies, potential climate change effects on mobility, aging populations, and terrorism reduction and mitigation. We cannot stop building infrastructure, but especially as many forms of the built environment outlast the context within which they were built—think of coal-fired power plants being built today that will be operating for decades—we can try to design infrastructure to be more resilient to change even when we cannot predict the particulars of such change, to be more adaptable and agile as those changes occur, and to be more aware of obvious effects of our designs that may have heretofore been ignored.

In a world where natural systems are for many purposes becoming simply another element of regional and global infrastructure systems, society in general and engineers in particular are challenged to rapidly develop tools, methods, and intellectual frameworks that enable reasoned responses. The reductionism inherent in pretending the challenge is just climate change or new transportation systems or lower-carbon energy production and distribution systems or water quality and quantity, although appropriate in certain cases, can no longer be defended as an acceptable disciplinary or institutional approach to infrastructure design, operation, and management. Rather, engineers, engineering educators, and institutions that support and manage infrastructure need to create solutions in the real world, and to embrace the implications of the Anthropocene. Bluntly, the world is a design space, and absent catastrophic collapse, there is no going back.

New Engineering for a Novel Planet

The reality of the terraformed Earth, whatever one wants to call it, requires some significant changes in the way technology and infrastructure are designed, implemented, understood, and managed. Although the obvious target disciplines are in engineering, such a focus would be inadequate. Increasingly, private firms develop and manage emerging technologies that shape the integrated human/built/natural infrastructures that are the defining characteristics of the Anthropocene, so business education writ broadly must be part of the shift. Moreover, as systems that were previously considered external to human design, such as biology and material cycles, increasingly become design products, many aspects of science segue into engineering fields, requiring different mental models and tools. This implies that many fields of science also require some grounding in technology, design, and engineering. Against this background, and recognizing that any suggestions for reform at today's early stage will necessarily be partial and somewhat arbitrary, we can nevertheless make some recommendations for change. For simplicity, we will group these suggestions into three categories: short term, medium term, and long term.

Short term. In all relevant disciplines, including at least all engineering and business programs, at least three areas of learning should be mandatory for all students, undergraduate and graduate, and professional accreditation institutions such as the Accreditation Board for Engineering and Technology (ABET), which approves engineering programs, should set an explicit timetable for achieving this. Whether these areas of expertise are introduced as modules in existing curricula or as independent courses is not as important as ensuring that some level of substantive knowledge is acquired by all students. In order to reinforce this, professional exams, such as the Fundamentals of Engineering, and Principles and Practice of Engineering exams given to professional engineers, should be expanded to include these subject areas. These broad areas of competence are:

1. Technological, social, and sustainable systems theory and principles, including concepts such as wicked complexity and satisficing versus optimizing system performance. This category of knowledge includes understanding the difference between simple and complex system behavior, and designing and managing integrated human/built/natural systems.

2. Big data/analytics/artificial intelligence (AI) functions and systems. Any business or engineering student who graduates without knowing something about these fields is verging on incompetent.

3. Cybersecurity and cyberwar operations from a defensive and offensive perspective. Today, business managers and engineers are trained to rely on, and design and deploy, "smart" systems at all scales; indeed, in many cases this is a performance and business imperative. On

the other hand, virtually none of these professionals are taught anything about cybersecurity or cyberwar, a gaping competency chasm when both Russian and Chinese strategic military doctrines have shifted toward deployment of unrestricted warfare / hybrid warfare / cyber warfare initiatives. Inviting an adversary to subvert all your critical infrastructure and systems, which the U.S. educational system now does, is simply stupid, especially since these adversaries are very engaged, and quite competent, in such activities.

These changes in curricula are a one-time improvement and by themselves will quickly become inadequate. At least as regards engineering, two further steps should immediately be taken. It is, for example, sadly the case today that too many engineering students are still being graduated with skill sets that prime them for replacement by expert systems. Thus, the first step should be to review the entire engineering curriculum for each discipline, identifying those elements of existing courses that software, AI systems, and changes in technology have made obsolete. A thorough analysis of the content of existing courses, coupled to modernization of ABET criteria, will help create educational programs that prepare students for the future, not the 1950s.

This should be combined with implementation of a change that engineering schools and those who hire engineers have been talking about for many years but have yet to implement: making engineering a graduate-level professional degree. Uniquely among professions, engineering remains, at least in theory, an undergraduate professional degree: neither law nor medicine nor any other major profession does so. This anachronism today works only by impoverishing the scope and scale of the education provided to undergraduate engineering students, who are then thrust into a world that demands ever more operational sophistication from them—something they cannot provide because there's no room for anything but engineering-related courses in their undergraduate programs.

Medium term. The Anthropocene is characterized by unpredictable change with far shorter time cycles than those common to most infrastructure systems. An important emphasis, therefore, must be on developing frameworks, tools, and methods that enable more agile, adaptable, and resilient infrastructure design and operation even in the face of fundamental uncertainty about virtually every important challenge infrastructure systems will face.

Although specifics are difficult and remain to be worked out, there are a number of ideas and models across existing engineering systems that can be generalized toward the goal of agile, adaptive, and resilient infrastructure. For example, the entire domain of consumer computational technology is characterized by many different technologies that must converge in working systems that are robust and simple enough in operation for average consumers to understand. One of the basic mechanisms for this process is "roadmapping," where modularity of technology—printers are different from data storage devices are different from input devices—is combined with robust module interfaces to enable both module and overall product evolution that is at once unpredictable, functional, and economic. Another example from consumer electronics is the substitution of software for hardware functionality whenever possible. Assuming appropriate hardware design, software can easily be upgraded, whereas hardware is far more difficult to change out (imagine if every improvement in software security or function required a new motherboard!). And programming offers a third model: the core of C++, for example, has remained fairly stable for years, but software evolution based on that core has been explosive, unpredictable, and very rapid. This suggests a "hub-and-spoke" framework: those elements of an infrastructure system that are not likely to change rapidly can be built with longer time frames and be more robust, whereas the spokes and edge elements, connected with stable interfaces, can evolve unpredictably and much more rapidly. Other characteristics might include planned obsolescence, resilience thinking that includes safe-to-fail and extensible design that can be easily updated, and increased compatibility and connectivity of hardware. These and similar models should be applied to infrastructure design generally.

Because of their importance to society, infrastructure systems tend to be designed to be robust and long-lasting. But especially as "natural" systems increasingly become a part of large infrastructure design and planning, it may be necessary to identify infrastructure where context may be changing rapidly, and "down-design" it: that is, design it to be less robust, more temporary, and capable of being replaced inexpensively and much more quickly. This could be combined with a major hubs/minor hubs/links design, where robust major hubs last for long times, minor hubs are down-designed, and structural links are robust in the face of unpredictable challenge.

Long term. The longer-term situation is more inchoate and more complex. Three fundamental shifts, one conceptual, one systemic, and one educational, are required, but none of them can be planned with any specificity. Accordingly, an overriding requirement is to learn how to implement continuous dialog and learning with the many systems within which we are embedded so that we can adjust our mental models, institutions, and educational practices in response to real time changes in system state. Policy and practice become processes of adaptive evolution rather than static responses to a stable world, and technocrats, managers, and engineers must learn to focus on ethical and responsible processes, and less on content, which will be continually and unpredictably changing.

The conceptual shift sounds simple. The fundamental implication of the Anthropocene is that everything from the planetary to the human body itself become design spaces: the world has become infrastructure. Especially given the complexity of the systems involved, this doesn't mean design in the sense of a completely understood controlled system such as a toaster. The world of simple, optimizable design is gone. Rather, it means that in a world of radical human life extension, integrated AI/human cognition, ecosystems as infrastructure, and re-wilding using previously extinct species in accord with wilderness chic design aesthetics, everything is subject to human intervention and intentionality. The Everglades is a design choice and exists today in its current state because we choose to have it exist in that state, even if it obviously has biological and ecological complexity beyond our current understanding. The simple truth of a terraformed planet is difficult for many people to accept, for reasons ranging from fear of the responsibility to ideological commitment to archaic visions of the sanctity of nature, but it is a necessary basis for technocratic and policy professionals if they wish to act ethically and responsibly under today's extremely challenging conditions.

Systemically, institutions need to evolve to mirror the complexity and interdisciplinarity of the networks with which they are interacting. Importantly, this means that institutions must become multi-ideological if they want to manage regional and planetary infrastructure. If they do not, they will be ineffectual. Thus, for example, the United Nations Framework Convention on Climate Change, signed in 1992, prioritized environmental ideology over other values, with the result that implementation has been sporadic, contentious, and ultimately unsuccessful. The shift involved here is also conceptual, and deceptively simple: institutions driven by activist stakeholder and dominant ideological demands must instead begin to frame themselves as problem-solvers, with the desired goal being not the advance of a particular agenda, but the creation of a politically and culturally stable solution to a particular challenge. Moreover, the institutions that typically govern infrastructure are mechanistic, characterized by hierarchical structure, centralized authority, large numbers of difficult to change rules and procedures, precise division of labor, narrow spans on control, and formal means of coordination. In contrast, industrial organizations in sectors exposed to rapid and unpredictable change, such as Silicon Valley technology firms, are far more organic, with the emphasis on moving fast rather than the lumbering pace imposed by formal structure. Whether current infrastructure institutions are capable of transitioning their management structures is an open question, given their legalistic and bureaucratic incentive structures; if they cannot, they must be replaced.

The educational shift is not unlike the institutional shift. Engineering education in particular combines quantitatively complicated but structured models and learning with creativity, design, operation, and problem-solving capability. The former is rapidly becoming mechanized, and the latter is increasingly done in a rapidly changing social, economic, and technological environment. Because of this, it will not be enough to simply reform engineering education in the short term: rather, engineering education must become a constantly evolving process and curriculum that changes at the same pace as its context. Engineers, who today often resemble expert systems—highly competent in one specific area such as chemical or civil engineering, but with knowledge that rapidly scales off at the edge of that domain because they don't have time to gain much education outside their particular discipline—must instead become constantly learning quantitative problem solvers, able to integrate across many different domains, as well as the still yawning gap between the science and technology fields and the social sciences and humanities.

These changes sound challenging, and indeed they are. But the Anthropocene is not a choice that we can reject; it is a reality that we have often unintentionally been creating since the Neolithic. We are now at the point where it must be embraced in all its complexity if we intend to respond rationally, ethically, and responsibly to its challenges.

Recommended Reading

B. R. Allenby, *The Theory and Practice of Sustainable Engineering* (Upper Saddle River, NJ: Prentice Hall, 2012).

B. R. Allenby, "Earth Systems Engineering and Management: A Manifesto," *Environmental Science & Technology* 41, no. 23 (2007): 7960–7965.

M. V. Chester and B. R. Allenby, "Toward Adaptive Infrastructure: Flexibility and Agility in a Non-Stationarity Age," *Sustainable and Resilient Infrastructure* 3 (2018): 1–15.

R. Costanza, R. de Groot, L. Braat, I. Kubiszewski, L. Fioramonti, P. Sutton, S. Farber, and M. Grasso, "Twenty Years of Ecosystem Services: How Far Have We Come and How Far Do We Still Need to Go?" *Ecosystem Services* 28 (2017): 1–16.

Y. Kim, D. A. Eisenberg, E. N. Bondank, M. V. Chester, G. Mascaro, and B. S. Underwood, "Fail-Safe and Safe-to-Fail Adaptation: Decision-Making for Urban Flooding Under Climate Change," *Climate Change* 145, nos. 3-4 (2017): 397–412.

B. L. Turner II, W. C. Clark, R. W. Kates, J. F. Richards, J. T. Mathews, and W. B. Meyer, eds., *The Earth as Transformed by Human Action* (Cambridge, UK: University of Cambridge Press, 1990).

4

FLEXIBILITY AND AGILITY IN A NON-STATIONARITY AGE*

Mikhail Chester and Braden Allenby

A 1963 quotation by management professor Leon Megginson—one that's often misattributed to Charles Darwin—states that "It is not the most intellectual of the species that survives; it is not the strongest that survives; but the species that survives is the one that is able best to adapt and adjust to the changing environment in which it finds itself."[1] The concept of adaptation and the complex principles that support it have been the focus of researchers in the fields of biology, but with significant application in other fields including business,[2] management,[3] and computer science.[4]

Adaptation is perhaps one of the most fundamental and powerful explanatory concepts for the changes in complex systems, in that it provides an explanation for the persistence of successful systems in the face of significant changes in internal and external environments. In biology,

---

* Reprinted from M. V. Chester and B. R. Allenby, "Toward Adaptive Infrastructure: Flexibility and Agility in a Non-Stationarity Age," *Sustainable and Resilient Infrastructure* 4, no. 4 (2018): 1–15, with permission from Taylor & Francis.

adaptation is a trait maintained by natural selection that enhances fitness and survival. More broadly, the concept characterizes the capability of organisms, complex systems, businesses, or institutions to change their organizing principles, structure, and behaviors to succeed in changing environments. How success is measured differs across disciplines, with biology focused on reproduction and business focused on maintaining growth and ultimately profitability.[5]

Yet when it comes to infrastructure, the systems that we've deployed and continue to maintain—the backbones of our cities, economies, and overall well-being—there appears to be limited capabilities to adapt. This raises serious questions about their ability to provide services in a future with changing demands, population, climate, security challenges, and environmental conditions.

The infrastructure that supports our societies provides untold benefits. Infrastructures are socio-technical systems composed of physical assets and the institutions that manage, govern, finance, and regulate them. The services they provide deliver resources such as energy, water, and information, and move and process waste. These services are not purely physical. While our focus on infrastructure is primarily on hard (or gray) systems—roads, buildings, power, water, etc.—we will also examine the role of soft (i.e., institutions) infrastructure and its relationship with hard systems.

Transportation infrastructure provides mobility and ultimately access to people, goods, and services. Buildings provide shelter for people, businesses, and services. Hard infrastructure can be characterized as services that either produce or deliver resources directly (such as energy, water, waste, or information) or provide mechanisms for resource consumption (such as buildings). In the United States, critical infrastructure is defined as chemical, commercial facilities, communications, critical manufacturing, dams, defense, emergency services, energy, financial, food and agriculture, government facilities, healthcare and public health, information technology, nuclear, transportation, water, and wastewater.[6] People's daily needs are typically met by municipal infrastructure and in many cases private energy companies. Supply chains for food, fuel, and materials are generally supported by institutions at larger scales—regional, state, and federal. As such, the funding and planning for adaptive infrastructure must recognize the client. More broadly, infrastructure facilitates derived demands; we don't usually demand the resource or service that infrastructure provides, but instead what that resource or service enables, in other words, the utility that it provides.

While the extent of infrastructure, what it delivers, and how it is used are somewhat quantifiable, the benefits of infrastructure ultimately are in its functioning as an engine of social well-being, which can be characterized through economic growth, health, quality of life, etc. To communicate the value of infrastructure, efforts have been made to monetize this well-being, both for gray infrastructure[7] and even ecological infrastructure.[8]

In the developed world, the core physical structures that define our infrastructure have often not changed in decades, sometimes centuries, from roads to water delivery to power generation and transmission. These infrastructures have certainly seen the implementation of new technologies (e.g., sensors and computing, automation, more efficient components) in support of the services delivered, but the core structures that have been used for decades (if not longer), from roadways to centralized fossil-based electricity generation to water distribution networks, are the cornerstones of the systems that we rely on today. Some are old and in need of rehabilitation or replacement. Some are new and likely to last a long time, making change difficult. And some are yet to be built, with the opportunity to affect their design.

Furthermore, infrastructure has often been built in support of the dominant technologies at the time it was conceived – not just in physical manifestations, but also in the rules, financing, and governing of the institutions that manage the infrastructure. This becomes a problem when the demands that we ask infrastructure to satisfy change and that infrastructure cannot change quickly enough to meet these demands.

In the past century we've seen the design of many technological and institutional forces that lock-in infrastructure systems. Prioritization of funds to roadways after World War II and minimum offstreet parking standards are perhaps the most dominant forces for automobile-centric transportation development in the United States.[9] The commitment of manufacturers through the investment of resources, labor, and manufacturing plants supported several waves of technological innovation that used polyphase electric supply.[10] The rapid growth of water utilities and distribution systems in the late 1800s was largely driven by concerns for public health (through the emergence of bacteriology) and safety, to protect growing populations against disease and fires. Centralized water distribution systems grew exponentially between 1880 and 1895 (from roughly 600 to 3,000 in the United States), while at the same time regulatory agencies emerged to ensure provision of services and affordability.[11]

As new technologies come online or as demand for services changes, our infrastructures (both hard and soft) may be unable to adapt, raising questions about how quickly they can change given new societal needs or threats. Given that our infrastructures tend to persist for long time periods, are they agile? Can they adapt to changing conditions? Why do our infrastructures need to be adaptable? Why and how should we design our infrastructure to be adaptable?

There are fundamental reasons why these questions arise at this time. There is a critical category distinction between physical infrastructure designed to be part of an overall infrastructure that is intended to last many decades, and the shorter and more abrupt changes in economic, technological, social, and institutional systems that are coupled to infrastructure. If the rate of change of these latter systems is relatively slow, as it has been for most of our history, infrastructure with half-lives of decades is not a problem.

If, however, the rate of change of the system accelerates, we reach a point where the cycle time of infrastructure change decouples from the increasingly rapid social systems which they serve. We have seen this happen in other long-lived institutions such as law.[12] For example, the shift to autonomous local vehicle service from owned automobiles is happening much faster than in historical periods, yet we have just begun to think about the implications for urban and transport infrastructure design.

Here we attempt to answer the aforementioned questions. We start by identifying several major challenges that have created a crisis for current infrastructure. We then attempt to unpack the design principles for infrastructure in the past century and how these principles constrain our ability to adapt infrastructure to challenges. We then characterize – based on evidence from industries that have successfully deployed adaptable infrastructure – the novel design and operation principles for infrastructure for a future in which demands on our systems are changing rapidly and there is heightened unpredictability across a number of domains.

Infrastructure Challenges in the Non-Stationarity Age

Infrastructure systems are facing several major challenges that threaten their performance, the services they deliver, and ultimately the well-being of the societies that rely on them. The confluence of these challenges can be described as a *crisis*. This is especially true in the United States, where significant attention is now focused on the state of disrepair of many major infrastructure systems,[13] but is also true in many other developed regions of the world. We posit that these challenges are 1) inflexibility; 2) funding; 3) maturation; 4) utilization; 5) interdependencies; 6) earth systems changes, most immediately climate change; 7) designing for social and environmental well-being; 8) transdisciplinary practices and processes; and 9) geopolitical security.

These challenges are interrelated and several produce so-called non-stationarity effects. We define non-stationarity loosely on statistical definitions as the unpredictability of future conditions based on past trends. There is a rich discourse around how climate change produces non-stationarity. For instance, precipitation and rates of discharges of rivers are becoming increasingly difficult to predict.[14]

We argue that funding for public infrastructure (namely transportation and water) also now exhibits non-stationarity. This is the result of policies (at federal, state, regional, and municipal levels) and financial planning that now inconsistently allocates funding (partly the result of escalating rehabilitation and maintenance costs) and creates significant uncertainty as to how much funding will be available for upkeep. This non-stationarity combined with the other challenges creates a crisis that must be imminently addressed to ensure that we are able to adapt infrastructure for the future.

1) *Inflexibility*: A unique characteristic of hard infrastructures and the soft infrastructures that support them is that they provide services for demands that are difficult to change except incrementally, even in the long-term. Some exceptions exist, notably information and communication technology (ICT) services, which have changed radically over short time periods. Inflexibility, which we'll explore in more detail later, emerges partly because physical infrastructures don't need to significantly change form as the services they've delivered have remained fairly consistent

for long periods of time. Unlike microchip fabrication, automobile manufacturing, and ICT services, the demands that infrastructure facilitate are relatively consistent on decadal scales. Electricity consumption, mobility (particularly by automobile), water use, and waste management demands are similar to demands 10, 30, 50, even 80 years ago. There have, of course, been efficiencies and technological improvements implemented in these infrastructure systems, and some, such as ICT, may change more rapidly than others, but their core physical structure has not changed dramatically in the long-term.

2) *Funding*: While there has been much attention focused on the state of disrepair of infrastructure in the United States, we argue that a major challenge is the sustainability of funds, particularly for long-term rehabilitation and technological improvement. Funding sustainability challenges result from two major forces: 1) many infrastructures were deployed in the middle of the last century and are now in need of major rehabilitation; and 2) there remains significant uncertainty about the availability of funds for this rehabilitation. The explosive growth of hard infrastructure in the United States with the New Deal, but more substantially after World War II, continued through the latter parts of the twentieth century. The Silent Generation, born between the 1920s and 1940s, experienced heavy capital investment (as a percentage of gross domestic product) in new infrastructure, which continued through the 1970s.[15]

As infrastructure built in the middle of the twentieth century began to reach the end of its service life, new pressures emerged to rehabilitate these systems. Many public agencies find themselves with insufficient funds to cover maintenance activities as these rehabilitation demands have grown.[16] Municipalities are forced to triage their limited rehabilitation funds, deciding which components of infrastructure get rehabilitated while delaying others.[17]

In the United States a compounding challenge is the uncertainty of federal funds. The Highway Trust Fund, for example, is supported by federal fuel taxes that have not been increased since 1993 and is not indexed to inflation.[18] Furthermore, many states use their fuel tax for purposes other than transportation.[19] There have been instances where the fund was projected to become insolvent. Major challenges also exist for water, electricity, aviation, waste management, rail, and other infrastructures.[20]

3) *Maturation*: Some infrastructure in developed regions of the world has grown to a point where substantial expansion no longer takes place. This infrastructure is mature in the sense that cumulative increases in physical infrastructure and its capacity have leveled off over time, and funding priorities are shifting from capital investment in new infrastructure to increased investment in maintenance and rehabilitation of existing infrastructure.[21]

This maturation can occur for several reasons. First, there are limits to the outward extent that infrastructure can grow, not necessarily physically, but more so practically. Resource, demand, and budget constraints, including travel-time budgets, natural boundaries (such as oceans, bays, mountains, and protected land), and growth boundaries, can limit how far outward people and services will be—and ultimately how infrastructure is deployed.[22] There are certainly many places where infrastructure is still being deployed outward; however, infrastructure grows where people demand services. If demographics, costs, or other barriers reduce or constrain that demand, then growth will slow or stop.

Related to maturation is geographic scale. In many countries, and particularly the United States, infrastructure exists on such large scales that meaningful and timely changes may require herculean efforts. Take for example the U.S. roadway, railway, power, and pipeline networks (Figure 1), which when visualized show the boundaries and urban centers of the country. Replacing or enhancing infrastructure at large scale means a big change in physical assets and across many geographies. Even if unlimited funds were available, physical resource, time, and manpower constraints likely exist, and non-technical barriers may be so large that the rate of change is severely limited.

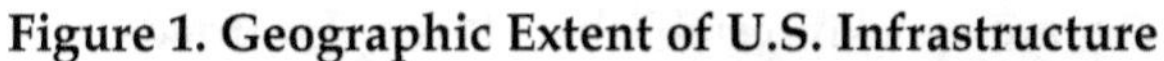

Figure 1. Geographic Extent of U.S. Infrastructure

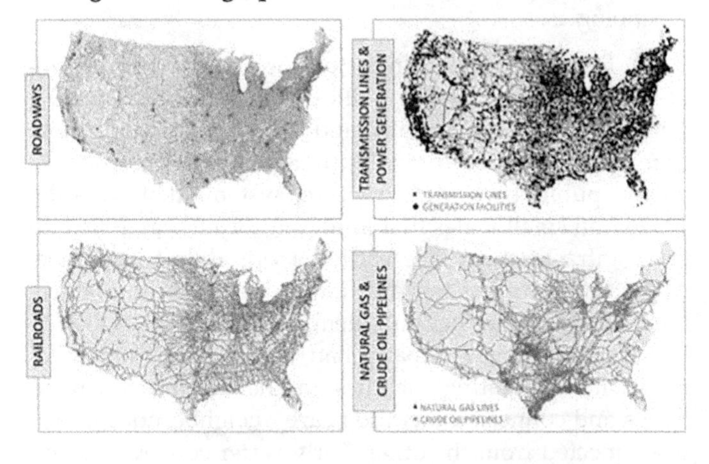

4) *Utilization*: Long-term infrastructure capacity planning remains a major challenge for financing and upkeep given the centralized nature of systems, lack of modularity, and resulting inflexibility. Infrastructure capacity is often planned on decadal scales, with forecasting of traffic demands, water consumption, and power consumption, for example, developed with increasingly sophisticated models. Yet accurate forecasting for roughly 30 or more years out remains elusive given the increasing uncertainty associated with the multitude of variables that drive infrastructure use: population, socioeconomics, climate, technologies, economics, and activities.

With largely centralized and inflexible infrastructure, managers will ultimately be confronted with the challenge of infrastructure that is either under- or oversized, sometimes grossly. This is evident in the oversizing of infrastructure after the population collapses of Detroit and New

Orleans, or undersizing in the cases of cities that have experienced rapid population growth, such as Phoenix and Las Vegas. While oversizing is apparent through cries for more funding to maintain underutilized systems or derelict structures, undersizing is not usually as obvious as short-term policies to meet, say, rapid changes in population growth are quickly established to deploy tried-and-true technologies.

5) *Interdependencies*: Infrastructures are becoming increasingly interdependent with other hard infrastructure, with managing institutions, and with information (more and more delivered digitally). Imagine early instances of shared public hard infrastructure, systems that were deployed on small scales that in no significant way relied on other infrastructure. The earliest roads didn't have electronic traffic control nor did they have power lines above them or water lines beneath them. By the late 1800s the Edison Illuminating Company had deployed a number of electricity generating facilities in the Northeast United States and connected them to nearby neighborhoods, each disconnected from the other. Early water conveyance and distribution systems exclusively relied on gravity. Electrical pumps didn't appear until the early 1900s.[23]

Today, vast and centralized infrastructure systems are deeply connected with each other. Infrastructure can be interdependent in several ways: geographic (co-location or in close proximity); physical (output of one system is an input into another); cyber (data or information from one system is input for another); and logical (the social, financial, political, etc. relationships between infrastructure).[24] Power, water, and ICT share space with roadways; pipelines sometimes follow rail rights-of-way; and critical systems are often found at the same spatial location, also known as a geographic interdependency.

Because investments in hard infrastructure are fixed, sunk, and irreversible, they are a large risk. Sharing (in terms of co-location or hardware) physical infrastructure can reduce the costs of entry, making it easier for new players or technologies to compete in a market.[25] Infrastructures that deliver resources (such as energy or water, and also ICT) serve as the backbone of other infrastructure. Traffic control, train propulsion, water pumping and treatment, and communications rely on electricity. Wet-cooled thermoelectric facilities rely on water systems. Virtually all infrastructure needs transportation services to move people and goods. And the digital age has shifted mechanical controls to digital and introduced remarkable opportunities for generating, transmitting, and processing digital information – processes that are now deeply embedded in many infrastructure processes (cyber interdependencies).

The degree to which this embrace of ICT has introduced new vulnerabilities and unpredictabilities into infrastructure systems is underappreciated and poorly addressed in most infrastructure systems.[26] The interface of hard infrastructure with the institutions that manage them produces logical interdependencies that define the rules, policies, and norms for how they are designed and operated, and how quickly they can change (more on this later).

These tightly coupled interdependencies are a challenge because they introduce complexity at scales and with outcomes that we poorly understand. A perturbation – or even worse, a failure – in one infrastructure can cascade to other infrastructures, leading to service interruptions. The complexity of these interconnected systems, the emergent behaviors of infrastructures when one is shocked, is largely unknown, and represents a critically important area of study when financial, security, or climate change disturbances are introduced.

6) *Earth Systems Changes, Including Climate Change*: There is increasing evidence that critical earth systems are becoming destabilized due to human activity. While climate change is receiving more and more attention, it is likely that we will need to manage other systems, including nitrogen, phosphorous, and water, going forward.[27] Infrastructure

creates a human-made world in place of a naturally evolving one, and at the decadal level the dynamics of changing Earth systems becomes important for engineers. Climate change is likely the most immediate and direct Earth systems hazard that we're confronting. As such, we focus on climate change as a case study that illustrates a fundamental challenge of infrastructure design and management.

Some weather-related extreme events are occurring with greater frequency and intensity,[28] and infrastructures, typically designed based on historical conditions, are vulnerable to both extreme and gradual perturbations. Infrastructures are the front line of defense against climate change. The services that they provide are critical during storms, heat, flooding, wildfires, and cold, in terms of the resources they deliver and their direct protection against exposure.

Infrastructures are typically designed against return periods, the frequency that the infrastructure will experience a particular intensity. For example, a bridge over a wash might be designed to maintain structural integrity for a 100-year return period, meaning a flow rate of water through the wash that is experienced on average every 100 years. Two major challenges exist. First, much of our existing infrastructure has been designed for return periods that are likely to change under climate change forecasts. A storm that has historically occurred at a particular intensity once every 100 years may now occur every 20 years.[29]

Second, codes require that designs be based on historical weather conditions that are no longer valid. Those who design infrastructure have not used climate forecasts and even if they were to, they would need different design processes that embrace the uncertainty associated with climate forecasts. We can expect indirect effects on infrastructure from climate change as well, including new conflicts, mass migration, and disease. How these effects will impact infrastructure remain largely unexplored. They nonetheless present serious risk to the reliability of infrastructure services and challenges for delivering services in a future marked by these events.

7) *Social and Environmental Awareness*: Gone are the days when infrastructure could be designed without serious considerations for social well-being and adverse environmental effects. The last half century has produced a mountain of knowledge about how the design, construction, and use of infrastructure affect people and the environment. Some of this knowledge has affected regulatory processes that require environmental assessments. In addition to the National Environmental Policy Act process requiring more and more disclosures through special-purpose laws, research on social equity and the rapid incorporation of sustainability principles has created novel thinking about how infrastructure should be deployed.

That's not to say that the deployment and use of infrastructure does not produce social and environmental impacts, but that we are more aware of these impacts and some measures have been put in place to reduce them. When deploying new infrastructure, the knowledge that has been generated from this past half-century of study is much more likely to be known by engineers, designers, and managers, as well as the general public, which is able to participate by voting, providing public comments, and protesting. While we are much more aware of social and environmental impacts, infrastructure designers and managers do not necessarily have the flexibility and resources to avoid them, and more broadly balance social, environmental, economic, and technical costs and benefits in a holistic but rigorous manner. Policies, financing, and codes may perpetuate existing practices despite evidence of negative outcomes.

8) *Transdisciplinary Practices and Processes*: Integration of disciplinary and institutional practices and processes is needed to reflect the interdependencies in not just physical infrastructure, but in the institutions and cultures within which they are embedded. As infrastructures have become

increasingly interconnected and as our knowledge of the complex systems in which these infrastructure function, provide services, and result in unintended tradeoffs grows, our traditional disciplinary boundaries are no longer sufficient. Infrastructures are typically designed, funded, and managed by a multitude of players, sometimes private and sometime public. They are governed and owned by different asset management systems, standards, businesses, and funding mechanisms. To effectively acknowledge and work within these complex arrangements, transdisciplinarity will be required.

9) *Geopolitical Security*: Several fundamental trends in geopolitical and military doctrine and strategy have come together to make security challenges a critical challenge to infrastructure. The first is a rise in non-state-actor violence, often in the guise of terrorism, against communities and societies. Because infrastructure systems are increasingly reliant on cyber for connectivity and software that can be hacked for operational capacity, deliberate attacks against infrastructure are ever more tempting for those seeking soft targets.

The second is a shift in military strategy by major adversaries of the United States and Europe, especially Russia and China, toward "hybrid warfare" and "unrestricted warfare," which reframe military confrontation as a conflict across all social and cultural systems, including infrastructure.[30] Along these lines, it is notable that at a particularly fraught moment in the Ukrainian-Russian conflict, Ukraine's electric system was hacked and taken down, substation by substation, on December 23, 2013.[31]

Finally, the extent of the Russian attack on American and European social and structural systems is just becoming apparent, and is far more significant than most professionals, embedded in their daily routine, realize. Indeed, a leading NATO analyst has voiced what many have concluded: "Recent Russian activities in the information domain would indicate that Russia already considers itself to be in a state of war."[32] No infrastructure design that isn't hardened against deliberate information attacks can be considered resilient; failure to design security into infrastructure from the beginning is a major source of fragility and vulnerability. And given that cities may have decades- or century-old infrastructure, there may need to be prioritization of assets when hardening.[33]

10) *Wicked Complexity*: Interdependent and even independent infrastructures are dominated by nonlinear interactions, emergent and self-organizing behavior, and distributed control: the key properties of complex systems.[34] These properties are defined by physical and non-physical factors and result in limitations on our ability to understand the emergent behavior of infrastructure systems, where the interactions at one level produce unanticipated phenomena at another.[35] And interdependencies explode this complexity.

Consider the 2003 North America blackout. What started as a single downed power line resulted in a cascading failure throughout the Northeast United States and Canada that left 55 million people without electricity, some for up to two weeks.[36] Beyond the power system, outages were experienced in the water, transportation, communications, and industrial systems.

Technical complexity results from several forces: accretion, interaction, and edge cases.[37] Accretion describes how infrastructures have accumulated and layered technologies over long time periods, to the point where it is no longer apparent how controls work (consider the use of 1980s IBM mainframes by the U.S. Federal Aviation Administration). The ease of interconnections coupled with accretion leads to interactions that over time and scale become so numerous that testing and understanding their behavior becomes challenging. Lastly, edge cases—exceptions to standard design and operating rules—introduce additional layers that obfuscate our ability to understand system behaviors.

Given the obdurate nature of infrastructures, their scale, and ubiquitous use, it can be argued that the systems that we so critically rely on naturally tend toward complexity. Complexity is not strictly the result of technical variables. The increased fragmentation of organizations that have some say in infrastructure, and the processes associated with accommodating different perspectives on how infrastructure is designed or managed has contributed to wicked complexity.[38]

How to build and operate infrastructure is a wicked problem. Wicked problems are characterized as 1) not understanding the problem until a solution is developed (implementation of infrastructure provides new insights into the problem); 2) having no stopping rule (once infrastructures are deployed you often continue to modify them based on changing needs); 3) having solutions that are not right or wrong (there are generally multiple ways to deploy infrastructure, e.g., route alternatives); 4) having novel problems (the multitude of technical and social considerations means that how infrastructure are deployed and operated for a particular circumstance are unique); 5) solutions are a one-shot operation (as per design theorist Horst Rittel: "One cannot build a freeway to see how it works"); and 6) having no alternative solutions (there may be no way to meet the need, or there may be many potential solutions but no single solution).[39] The combination of these factors means that infrastructures are wicked complex systems. It has become extremely difficult, if not impossible, both to predict how systems will behave across space and time when perturbations occur and to change systems toward future goals.

As we've transitioned into the twenty-first century, we will likely find our infrastructures increasingly defined by these challenges. Several of these challenges manifested during the latter part of the last century and combined with emerging challenges—specifically the non-stationarity introduced by climate change and financing—meaning that new models of infrastructure design, construction, operation, and use will be needed.

As services and technologies change, the demands that we place on infrastructure will also change. To meet these changing demands, infrastructure will need to be *agile* (in the face of both predictable and unpredictable challenges) and *flexible,* preconditions for *adaptability*. Yet when it comes to hard infrastructure we have not seen a system that has these characteristics. In the following sections we explore systems that have successfully implemented these characteristics to meet rapidly changing demands and environmental conditions. We identify the design principles and operating conditions that enable these systems to behave with these characteristics and discuss the changes that are needed in hard infrastructure systems so that they can meet rapidly changing demands in the twenty-first century.

Designing and Planning Principles

Successful infrastructure in the twenty-first century will need radically different design principles. Engineers will need to be part of a process that reconceptualizes infrastructure from the purely physical to a system that includes institutional components and knowledge as integral parts. Infrastructures will need to change their structure, behavior, or resource use as demands change. In doing so they will need to support the rapid deployment and growth of nascent technologies (such as renewable electricity generation, microgrids, gray water systems, material reuse, and autonomous and electric vehicles) as well as technologies that we haven't yet begun to envision. These technologies are liable to change not just the physical operation, but the mathematics, of the underlying systems in unpredictable ways.

A deeper challenge is that we're not just operating on the level of infrastructure itself, but at the implicit models of operation that we sometimes haven't revisited in decades, requiring us to expand how we think about the institutional and disciplinary ways that we think about infrastructure. They will need to create opportunities for embedding digital sensing, data processing, and data analysis (so-called "smart") technologies that improve our understanding of how interdependent and complex infrastructures behave; provide us with protective measures against vulnerabilities and added security; and enable a new understanding of how our built environment functions to improve well-being.

These infrastructures will need to be able to meet these characteristics with unpredictability around financing and in the face of both extreme events and gradual changes in climate and the challenges that result from climate instability—itself a proxy for the challenge of designing and building infrastructures in a world where human activities increasingly impact all environmental systems. And they will usher in new metrics of infrastructure success that measure the ability to meet rapidly changing needs and respond to perturbations. This will require a fundamentally new paradigm in how we design, build, and operate our infrastructure.

Flexibility and *agility* will need to be at the core of this new paradigm. In the context of hard infrastructure, we distinguish between flexibility and agility based on changing demands and non-stationarity. With rapid changes in technology and ultimately the services that our systems provide, infrastructure will need to be flexible to changing demands. Infrastructure will also need to be agile in that its physical structure, along with the rules, policies, norms, and actors who manage and operate it, will need to be able to maintain function in a non-stationarity future. This includes planning and responding to unpredictable events, such as extreme weather or budget shortfalls, disease events, security challenges such as cyberattacks and physical terrorism, population migrations, and other phenomena.

Infrastructure managers can expect that at some point in the future these events will occur but cannot easily or accurately predict when or to what extent, or for that matter how perturbations in underlying human and natural systems will manifest themselves. The combination of flexible and agile design and operation characteristics are the preconditions for *adaptability*. These challenges are monumental but they are not without precedent. Successes in implementing flexible and adaptable infrastructure for rapidly changing demands in other industries can offer invaluable insight into the processes that need to shift the design paradigm of civil systems.

How do we design our infrastructure to be adaptable? We don't know the ultimate forms of infrastructure that enable flexibility and agility, as it is likely that these forms have not yet been identified, developed, tested, or implemented. Also, we're not training engineers and planners to function in the integrated infrastructure systems of the future.[40]

However, we can identify the characteristics of flexibility and agility that have been successfully implemented in other infrastructures and their processes, and describe how they may translate to civil systems. These characteristics do not strictly belong to the physical world; they must also exist in the organizations and institutions that manage and govern physical systems. We synthesize characteristics of flexibility and agility from several industries that have successfully changed their organizations and physical processes to meet rapidly changing demands and respond to unpredictability. We focus largely on ICT and manufacturing. We also explore the shifting of technological functions within automobiles and automobile travel to characterize the substitution of functionality from physical to digital and efficiency gains within technological and infrastructural constraints. These industries have infrastructures that have evolved in response to economic and competitive pressures that are not usually felt strongly by either the professionals that design infrastructure or the public planners and managers who operate it.

Through this review we identify several characteristics that enable flexibility and agility. Given the rapidly changing demands for services that commercial sectors must often meet and the structure of public institutions that typically manage civil infrastructure, we question where best practices are most likely to arise. Finally, we attempt to organize these characteristics into a structured framework, identifying drivers and characteristics that produce competencies for flexibility and adaptability. This framework is shown in Figure 2.

Figure 2. Stimuli, Properties, and Competencies for Adaptive Capacity

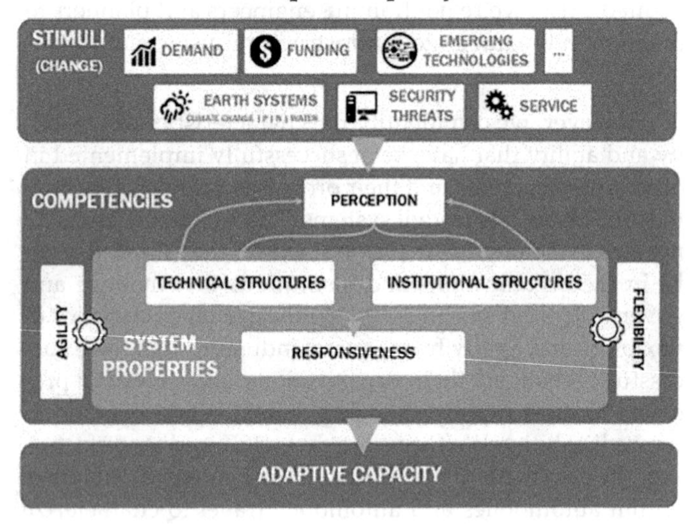

An adaptive infrastructure is one that has the capacity to perceive and respond to perturbations in such a way as to maintain fitness over time. Adaptive infrastructures have the capacity to recognize that stimuli or changes in demand are occurring or will occur, including the effects of these stimuli, and have the socio-technical structures in place to change quickly enough to meet future demands.

Stimuli can take many forms; we focus on those related to the aforementioned challenges. They describe a direct stress (e.g., climate change, extreme events, or inadequate funding for maintenance), change in demand (e.g., a rapid change in population resulting in more or less need for infrastructure services), or change in service (e.g., a technology or behavioral shift that brings about less-to-no need for an infrastructure). Other stimuli also exist, including emerging technologies and physical and cyber threats.

Competencies for adaptive capacity include agility and flexibility, but are preceded by the ability to perceive stimuli and how they will affect the system and ultimately the infrastructure. There is a rich body of study on how individuals and organizations perceive risk.[41] Infrastructure managers must be able to recognize that stimuli are or will occur and understand how they will affect the system. Beyond supervisory control and data acquisition systems, they must have physical and informational sensing capabilities that provide insight into the behaviors of complex and interconnected systems (both in terms of the infrastructure and its use). This is related to sensing and anticipating in resilience frameworks.[42]

Knowledge is a critical aspect of perception, which we argue is currently insufficient to deal with the aforementioned challenges facing infrastructure. As such, the capacity for infrastructure managers to anticipate the effects of stimuli is lacking. The capacity to perceive is a function of the technical and institutional structures supporting infrastructure,[43] and frequently fails when rates of change, or system complexity, exceed normal bounds. Institutions tasked with designing, managing, building, and maintaining infrastructure do so based on standard practices, codes, and methodologies reinforced by disciplinary expertise, training, and organizational culture.[44] In periods of rapid, non-incremental change, this disciplinary training produces barriers to the knowledge needed to understand and respond to stimuli, to perceive, to maintain fitness.

While infrastructure can take on many forms defined by network typology, public to private management, and national to local scale, the *system properties* that define physical configuration combined with the rules and objectives of the managing institutions ultimately affect its ability to respond. Responsiveness is defined as the propensity for purposeful and timely behavior change in the presence of stimuli.[45] The definitions of responsiveness, flexibility, and agility are often conflated, and there have been few efforts to differentiate. Following research by Ednilson Santos Bernardes and Mark D. Hanna, in the context of infrastructure adaptation we differentiate "responsiveness" as being the propensity for timely behavior change and "agility and flexibility" as being associated with reconfiguration of the system.

Competencies and appropriate system properties enable *adaptive capacity*, the ability of infrastructure to respond to inevitable and unexpected stimuli. Adaptive capacity has largely been defined by sociological-ecological systems researchers.[46] The dominant approach for designing infrastructure systems is that of risk management, that is, sizing infrastructure to be able to withstand an event of a particular magnitude and frequency (the 100-year return period is often chosen). This approach leads to large gray infrastructure that favor designs that keep hazards away (e.g., levees) or can continue operating during the hazard. Yet as infrastructure becomes larger and more permanent, the consequences of failure increase.[47] Furthermore, infrastructure becomes less adaptive when it has more legacy components that impede efficient system evolution in response to unanticipated stresses.

This approach, while robust in protecting against particular shocks, ultimately is highly inflexible for changes in demand beyond what was forecast; it can result in major consequences when failure occurs; and it is generally unable to cope with unforeseen stimuli. Adaptive capacity approaches are inimical to risk-based approaches in that they focus on maintaining capacity in the face of stimuli; minimize the consequences of stimuli instead of minimizing the probability of damage; privilege the use of solutions that maintain and enhance services; design autonomous management schemes instead of hierarchical; and encourage interdisciplinary collaboration and communication.[48]

We contend that the competencies and system properties needed to achieve adaptive capacity will require transformational shifts in the way that we build, operate, and perceive system purpose and function, and the educational and organizational institutions that we have historically relied on to design and manage infrastructure systems.

Adaptive Infrastructure

The competencies and system properties that can help enable adaptive capacities require novel planning techniques, technical and institutional structures, and integration of education and interdisciplinary practices across the lifecycle of infrastructure. Drawing on past successes from other industries, recommendations can be made for civil infrastructure systems. Following from Figure 2, the competencies and system properties associated with current infrastructure are contrasted with those of successful adaptive systems, based on the following discussion. These are summarized in Table 1 and discussed in detail as follows.

Table 1. Competencies and System Properties for Adaptive Systems

Competency	Current	Adaptive
Perception & responsiveness	Prioritizes perpetuation of existing designs	Roadmapping
Perception, responsiveness, & technical structure	Obdurate design	Design for obsolescence
Technical structure	Hardware-focused	Software-focused
Technical & institutional structures	Risk-based	Resilience-based
Technical structure	Incompatibility	Compatibility
Technical structure	Disconnected	Connectivity
Technical structure	Non-modular design	Modularity
Institutional structure	Mechanistic	Organic
Institutional structure	Culture of status quo	Culture of change
Perception & responsiveness	Discipline-focused education	Transdisciplinary education

For each of the four driving competencies identified in Figure 2, the current approaches and exploration of adaptive approaches are shown.

Roadmapping and Planned Obsolescence

Industry roadmaps have proven to be valuable for enabling radical innovation and evolution by aligning common goals across a number of different domains. We use the term "roadmap" to describe the development of a model or structure that allows multiple organizations with competing goals operating at many different levels of a technology system to plan together to enable the rapid evolution of systems and manage uncertainty.

In the 1990s during the early stages of development of much of today's ICT backbone and associated computational technologies and tools, roadmapping emerged as a valuable process for prioritizing technologies and identifying infrastructure gaps, suggesting robust interconnections between modules within which powerful innovation was occurring, and creating standards and business practices to meet these challenges across many (often competing) organizations.[49] For example, electronic industry roadmaps enabled constant improvement in computer performance at the user level even as disruptive innovation characterized component subsystems (e.g., portable storage devices evolved from large floppy disks to hard disks to thumb drives).

At the institutional level, roadmapping includes creating industry committees and associations, which subsequently use conferences, workshops, less formal collaborative practices, and other activities to create institutional structures that support constant innovation and communication in often highly competitive environments with significant antitrust and other legal constraints. Given that right now funding is often spent on expensive failures instead of preventative maintenance, roadmapping, when using measures such as return-on-investment, could help identify lower-cost pathways in addition to necessary technological change. Without roadmapping, the combination of unpredictable and disruptive innovation with smooth system-level evolution that characterized the growth of electronics and communications technologies across the entire spectrum of the ICT sector would have been impossible.

These roadmapping techniques suggest several general principles supporting agile and adaptive design. Most importantly, at an institutional level they suggest that a complex combination of competition, innovation, and collaboration can be managed through sophisticated use of modular design. They also suggest that rapid cycles of innovation and obsolescence within modules cannot just be tolerated, but encouraged, even as the overall system remains stable. In ICT, they helped to enable rapid and unpredictable development in ICT subsystems while maintaining a framework that ensured the overall technologies remained operational.[50] They managed unpredictable and important innovation within the module, yet maintained interconnection at the systems level.

Applying a roadmap model to infrastructure becomes a way to enable radical innovation within systems, while supporting a high level of inter-domain communication and constantly improving product, providing continuous and uninterrupted functionality to users. Consortia including cities and firms can generate such a roadmap that not only crosses engineering domains (e.g., energy, water, ICT, and transport, at a minimum), but policy domains (e.g., tax policy and transportation management), and institutional domains (e.g., the city government with all its silos and the critical private firms in each sector) to facilitate the planning and operation of next-generation infrastructure in non-stationarity conditions, and to encourage continuing innovation and efficiency in provision of services without disruption and at low lifecycle cost.

Roadmapping can be valuable for shifting design considerations from obdurate paradigms to planned obsolescence. For many components of infrastructure systems, managers favor designs and assets that can last a long time. This paradigm is problematic in that it locks users into technologies for the long-term, constraining the ability to modernize systems for changes in demand. *Obduracy* in infrastructure persists because managing institutions are constrained in their ways of thinking and changing one component requires multiple changes; it is therefore easier to maintain old configurations than introduce innovation.[51]

Alternatively, *planned obsolescence*, that is, planning for changes regarding function, profitability, and other dimensions of performance, can result in greater capacity to substitute infrastructure components and technologies to more efficiently meet changes in demands.[52] Access to infrastructure is important when replacing hardware. Much of our infrastructure is buried underground where access is costly and conditions are often unknown. This inaccessibility encourages waiting for failure. Accessibility will be critical for ensuring quick and frequent replacement or upgrades. While infrastructure obsolescence mostly has been studied from a physical asset perspective, the integration of computing and substitution of software for hardware function creates new opportunities for shifting functions.

Software-for-Hardware Substitution

With the increasing availability of sensors, processors, and data analytical tools at decreasing costs, there is a growing substitution of software for hardware, transitioning physical to digital processes that increase the flexibility and adaptive capacity of technologies while improving their efficiencies. These technologies are sometimes collectively referred to as "smart" technologies or systems. They are increasingly replacing or being used to augment the capabilities of physical processes. With their use, industries are finding that core business practices can be shifted.[53]

Sensors can predict the structural health of hardware, notifying operators of needed maintenance,[54] measure fluctuations in manufacturing processes, and adjust inputs to improve production and reduce costs.[55] They can also provide real-time information to users or software to adjust operations, likely avoiding the need for manual labor and associated resources. Fuel sensors in automobiles can adjust the air-to-fuel ratio, optimizing combustion and emissions. Digital technologies now allow industrial manufacturers to configure processes virtually before changing or upgrading equipment to proactively identify potential incompatibilities in parts or processes.[56] Variable frequency drives that use electronics to monitor motor performance and load requirements to optimize work by pumps, eliminating the need for smaller pumps and control valves in applications such as water distribution.[57] Traffic camera software is now smart enough to identify cars, pedestrians, and bicycles, reducing the need for in-pavement loop detectors and the associated asphalt impacts.[58] Implementing wireless communications instead of landlines reduces the need for wiring.

In addition to the efficiencies that are gained in substitution, there is likely less waste when you upgrade via software instead of hardware. Additionally, software-driven functions are more agile when faced with unpredictable and rapid change than hardware-driven functions that require physical alterations for upgrades. As software has progressed, the prior practice of upgrading via physical media—sending a disk with the upgrade through snail mail, for example—has become largely obsolete, replaced by the use of online software fixes that are more efficient and produce less waste. This is an important evolution, improving the agility of embedded software systems. Real-time fixes are necessary in an environment where malware and viruses are instantaneous; the next step is to have artificial intelligence responding to challenges and changes as they occur on a network-wide basis.

The implementation of smart technologies within existing infrastructure and technologies can create efficiencies within the constraints of inflexible systems. They can also help improve our understanding of the increasing complexities of infrastructure. Take, for example, the use of GPS, smart phones, and navigation software in personal automobile travel. Initially the automobile consisted of mechanical components with little to no interaction with the environment except through the driver's decisions. With the advent of electronics, automobile systems began working together, communicating information to each subsystem so that subsystem-specific adjustments could be made, thereby increasing the efficiency of the overall vehicle. Currently smart technologies are introducing new efficiencies.

Within the confines and rules of the roadway system, sensing technologies and software are now able to communicate to drivers the shortest paths, including routes that avoid traffic, thereby possibly saving fuel and time.[59]

These technologies introduce agility within the constraints of the current infrastructure, which is now up to a century old. There are of course limitations to the benefits that these technologies bring. The inefficiencies or poor condition of old infrastructure may prove to be a limiting factor in how much improvement smart technologies can provide. In the future it is conceivable that smart technologies will know the condition of infrastructure and reroute flows or traffic away from vulnerable links, or prevent component failures from cascading through or across infrastructure. They may provide us with insight into the increasing complexities of infrastructure.

Another benefit of hardware-to-software substitution is the integration of modules into a larger system or with the environment, effectively changing the scale and scope of the system that can then be optimized. With a mechanical system, efficiency improvements are largely confined to particular modules. Prior to the integration of software and smart technologies into automobiles, it wasn't possible to optimize the performance of the vehicle in real time, much less reach beyond it. But with sensors and ICT, you can not only optimize the automobile in real time, but make it part of a much larger system that includes other vehicles, infrastructure conditions, and traffic controls. This hardware-to-software substitution shifts not only the underlying technologies, but also the larger governing rules of travel; today navigation software can provide improvements across the entire transportation system and onboard software can optimize how your vehicle is performing in real time.

Risk-to-Resilience Thinking

Infrastructure will need to operate in natural environments defined by non-stationarity. For example, infrastructure systems are already experiencing more frequent extreme weather events, raising questions of whether traditional risk-based approaches to design are adequate. Many infrastructures (or their components) are designed for specific return periods; for example, a 100-year precipitation event that characterizes, in this case, an event with a particular intensity that has a 1% chance of occurring annually.

However, there is growing evidence that these events will become more frequent and unpredictable, raising questions of which return periods to design to, the affordability of larger designs, and whether people want to live near bigger structures. This risk-based approach, which focuses on the risk triplet of *threats* × *threat probability* × *consequences*,[60] is often based on historical data and results in large gray infrastructure with low probabilities of failure, long lifetimes, and oversizing (e.g., levees or retention basins).

The problem is that the risk management approach does not incorporate an understanding of what may happen when the infrastructure itself fails. Larger and more permanent infrastructures tend to be associated with greater damages when they fail.[61] Climate change necessitates new approaches to infrastructure design—approaches that recognize risks but are adaptable, in that they do not compromise the entire system upon failure. More broadly, infrastructure will need to be designed with new rules that recognize the non-stationarity in Earth systems created by human activities.

Compatibility, Connectivity, and Modularity

ICT, particularly those technologies developed in the internet age, are designed to meet rapid changes in demand

and types of services offered. Early work in ICT recognized the necessity of flexibility as both a system attribute and core competency.[62] The characteristics of flexibility are defined as compatibility, connectivity, and modularity.[63]

Compatibility is the ability to share information across different technological components, involving integration rules and access standards that affect shareability and reusability. *Connectivity* is the ability of any technology to communicate with components inside and outside of the system. It is a measure of the number of processes that are able to interact. Connectivity enables shareability, which is central to flexibility in that it allows resources to be used for new functions.[64]

Modularization in the manufacturing and computing industries has helped manage complexity, enabled parallel work, and accommodated future uncertainty.[65] *Modularity* is the ability to add, modify, or remove components easily, without needing to change other modules and subsystems, achieved through standardization.[66] Integrated systems that lack modularity have fixed processes embedded within the structure that interact, but cannot be easily removed or reconfigured. Likewise, these systems are not capable of easily adding new processes. Systems become more manageable when processes are modularized—that is, processes are designed with standards for information and hardware interactions, and the core routines are compartmentalized. They can be designed independently and used in a variety of situations or systems. Modern computing uses modular design in both software and hardware to enable rapid responses to changing customer demand and manufacturing processes more adaptive.[67]

The explosive reach of the internet both in terms of information and hardware embodies these three characteristics. Standards for information transfer, such as web languages (e.g., HTML and FTP), transmission and information control protocols (i.e., TCP/IP), and the heavily modularized use of hardware and software, have enabled the internet to grow at a pace never before seen with other technologies.[68]

Flexible Management

Successful organizations reflect the complexity of the environment in which they operate.[69] Contingency theories state that organizations must be analyzed as open systems that directly interact with their environment, and as such, in order to be effective, must be able to adapt to changing contingencies.[70] Organizations that can successfully operate in unstable, changing, and unpredictable environments have organic design characterized by less precise division of labor, wider span of control, more decentralized authority, fewer rules and procedures, and more personal means of coordination.[71]

This is in contrast with the typical mechanistic design of organizations that manage infrastructure. They are characterized by highly hierarchical structures, formal management with a centralized authority, a large number of rules and procedures, precise division of labor, narrow span of control, and formal means of coordination. Table 2 contrasts these characteristics in the context of infrastructure.

The mechanistic form persists because of historically relatively stable and predictable demands and environments. The ways in which water, electricity, and mobility are demanded have not changed significantly in the past century. The mechanistic approach has been shown to be most effective in environments that require routine operation and little change. In these environments, high-level management possesses the appropriate amount of knowledge to make decisions and organize work.

However, when the environment becomes unstable, management cannot acquire all the knowledge associated with the changing environment, and distributing the knowledge and decision making at the bottom of the hierarchy becomes more effective.[72] This is because in order for one system to be able to understand and manage another, it needs to be of the same or greater complexity.[73] Organic structures allow for more internal specialization to respond to changing environments, thereby increasing responsiveness.[74]

The scholar Bohdana Sherehiy, citing a large body of literature, argues that flexible and adaptable organizations have fewer regulations of job description, work schedules, organization policies, and power differentials (e.g., titles). They have fewer levels of hierarchy, informal and changing lines of authority, open and informal communication, loose boundaries among function and units, distributed decision making, and fluid role definitions. Furthermore, authority is tied to tasks instead of positions, and shifts as task shifts.[75]

Table 2. Characteristics of Mechanistic and Organic Infrastructure Management Structures

	Current (Mechanistic)	Adaptive (Organic)
Authority	Hierarchy	Less adherence to authority and control
Communication	Hierarchical	Networked
Knowledge	Centralized	Decentralized
Loyalty	Organization	Project
	High degree of formality	High degree of flexibility and discretion
Coordination	Formal and impersonal	Informal and personal
Rules and Procedures	Many	Few
Tasks	Specialized	Contribution to common tasks

Adapted from B. Sherehiy, W. Karwowski, and J. K. Layer, "A review of enterprise agility: Concepts, frameworks, and attributes," International Journal of Industrial Ergonomics *37, no. 5 (2007): 445–460.*

Significant questions remain as to whether the public institutions that manage infrastructure have the flexibility to change from mechanistic to organic cultures: do laws and policies constrain their organizational form? Given that public institutions are directly beholden to taxpayers, can they take on new organizational forms or change how infrastructure performance is measured? This should not be taken as an argument that infrastructure should be privatized, but instead as a challenge to conceptualize new management structures that embrace organic characteristics.

Taxonomies of organizational flexibility tend to focus on a few key factors to meet changing demands, which may be helpful when developing new organizational structures for managing infrastructure.[76] They alter the number of employees and hours through employing part-time, temporary, or short-term contracts, or by changing working times. They create opportunities for changing workforce skills to accomplish a wider range of tasks. And they recommend financial flexibility through pay-for-performance and profit-sharing plans.[77] Whether these factors are feasible for public institutions in the context of infrastructure management remains an open question.

Culture of Change

Related to flexibility in management is a culture of change, an organization supportive of experimentation, learning, and innovation, and one that is aware of changes in the environment.[78] Whether infrastructure organizations are capable of achieving these remains an open question as there are a lack of incentives (related to what has historically been consistent market demand), legal and regulatory requirements (such as adherence to codes and regulations), safety requirements, reliability requirements, and constraints (e.g., missions that are tied to public goods, or earmarked funding).

Given that infrastructure is typically associated with the provision of public goods, many constraints exist to ensure that resources are delivered reliably, fairly, and at the lowest cost. These constraints may be inimical to rapid experimentation, innovation, and change. We can even question whether it makes any sense to reimagine public institutions that manage infrastructure as reflections of private enterprises that must persist in very different environments.

However, there remains a pressing need for infrastructure organizations to be able to change to ensure reliability and foster new activities and technologies into the future. Change-oriented cultures are partly the result of education, which can take several forms. Organizations can directly support research activities and structurally integrate research into decision making at all levels. They can budget for experimentation or the testing of new and emerging infrastructure designs and management strategies. They can consider changing organizational and individual responsibilities toward anticipated stressors (and ultimately solutions) and away from structures that are purely disciplinary and focused on process or challenges within those disciplines—toward competencies that span multiple disciplines. Part of this responsibility falls to those educating the next generation.

Education

Institutional design is partially an artifact of training, and infrastructure education (largely engineering) continues to emphasize knowledge and problem solving within single domains. Relatedly, infrastructure continues to be planned, designed, and operated as rigid silos with little to no understanding of the complexity that emerges from the inherent interdependencies of systems. The result is that each system, to the extent it does try to optimize, does so within subsystems.

Integrated education at the university level, and integrated planning in practice, is almost the opposite of what educators and engineers do today. But it's necessary for understanding the impacts of stressors and opportunities for developing strategies to handle stressors going forward. The true complexity of mitigating the infrastructure crisis includes challenges not only in physical infrastructure and the institutions that manage them but at all levels of knowledge production, starting with education and training. Engineers must be able to think about institutional design in these complex environments, as well as technical design. And engineers don't necessarily need all of the expertise themselves; they could be part of teams that design, deploy, and operate systems.

Lock-in and Path Dependency

Transitions toward agile and flexible infrastructure will require the identification of, and strategies to overcome, barriers that perpetuate current infrastructure forms despite a need to change. These barriers will need to be catalogued and associated with the actors and forces (rules, policies, norms, financing, etc.) that support them. There is a long history of describing how major changes are needed for infrastructure as well as the technologies that use these infrastructure, often expressed as scenario analyses showing how things can be given some monumental aligning of forces.[79]

Although some infrastructures have changed quickly (e.g., the shift from landlines to wireless technologies in ICT), when it comes to many core civil systems (i.e., water, electricity, transport) large-scale transitions have not happened. This is because many barriers exist that prevent these transitions, including financial (lack of funding for capital investments or earmarking of funds for particular purposes), political (limited political will and "not in my term of office" mentality), codes (such as minimum parking requirements), social (communities may not see the value of redirecting resources from an established technology to an alternative), cultural (for example, consumers' unwillingness to consume treated wastewater), and technological forces.

We describe *lock-in* as the inability to change infrastructure due to these barriers and their often-synergistic interactions with other infrastructures (i.e., interdependencies where we cannot radically change the structure or function of one infrastructure because another relies on that structure or function). Many infrastructure designs have persisted for so long that other infrastructure and institutions have become interdependent, leading to additional barriers and complications for making transitions.

Furthermore, given that demand for infrastructure services doesn't change quickly, managers and policymakers end up prioritizing low-cost rehabilitation and supporting established technologies. Despite the dire state of infrastructure,[80] as long as people get basic services cheaply, the impetus for major reform will not exist. New funds tend to go to existing infrastructures and technologies because we know how to do them inexpensively and we've codified so much for them. Also, regulated infrastructure is often permitted to expend capital for variable costs, but not for fixed costs or capital improvements; American railroads are a case in point.[81] While the American Society of Civil Engineers has done a good job of communicating the state of infrastructure to policymakers and engineers, they do not engage directly with the general public—the group that will need to pay to avoid aging and failing infrastructure.

The persistence of forces that maintain lock-in into the long term creates *path dependency*, a characteristic of all complex adaptive systems whereby past conditions significantly impact possible future trajectories. In the case of long-lived infrastructure in a period of rapid change, path dependency can lead to the perpetuation of infrastructure and the technologies, activities, and behaviors that rely on them, despite alternative futures being preferable. These alternative paths may describe futures with lower user and public agency costs, reduced energy use and greenhouse gas emissions, or more socially equitable outcomes.

As infrastructures and the technologies they support continue into the long term and interdependencies are established with other systems, there exists the possibility that the achievement of benefits becomes limited. Learning how to design infrastructure in such a way as to enable more desirable future states given the reality and complexity of path dependency thus becomes a necessary future competency.

We characterize the confluence of constraints (barriers) that prevent us from achieving a more desirable state as a *limit*. In mathematics, limits are values that functions approach as the inputs approach some value. In the case of infrastructure, limits can be thought of as the practical achievable futures given technical and nontechnical barriers. The concept of a limit is important because it recognizes the reality that the infrastructure that we've integrated into all facets of society may not allow us to reach a desirable future, that there's only so much better we can make it or the activities that use it.

For example, given the preference and incentivizes for automobile-focused roadway infrastructure, there may not be a pathway where greenhouse gas emissions can be reduced fast enough to avoid some significant climate change threshold. Or consider that the design and deployment of housing stock (specifically the materials, building technologies, and form) may limit our ability to reduce building energy use beyond a certain threshold.[82] As such, our current infrastructures and the forces that maintain their persistence are limiting our ability to achieve desired goals.

More subtly, so is our educational system. Educating engineers to think of systems only in terms of traditional technological frameworks, such as energy, transportation, or information, makes it difficult to achieve integrated design and management of infrastructure, and thus impedes the ability to reduce the deleterious effects of lock-in. And yet engineering education has its own path dependency, reflecting such constraints as meeting requirements for professional and school accreditation.

Understanding how lock-in with various coupled systems is acting in a particular design space can help identify the barriers that need to be overcome to create opportunities for agile and flexible infrastructure. There is a long history of scenario analysis that shows that it is technologically feasible to reach more preferable future states. Research is needed that identifies the barriers (financial, political, social, technological, institutional, educational, etc.) to achieving these states and competencies needed for overcoming these barriers. These competencies will likely center on changing social and cultural norms—and current educational practices and institutions—that drive the formalization of codes and financial structures, and the institutional practices that drive infrastructure growth and management.

A Century of Change

A century from now we may look back and ask ourselves what the costs were of not transitioning our antiquated infrastructure fast enough. These costs would have taken many different forms, from missing economic benefits to environmental losses to possibly even loss of life (consider infrastructure not sufficiently protecting us from climate change).

Or maybe we will look back at the monumental success that was the rapid transition of physical infrastructures and their managing institutions, and the accumulating benefits that it afforded. In this future, pipes may still deliver water, lines may still deliver power, and roads may still move people and goods, but infrastructures (as socio-technical systems) will have the competencies and structure to be able to meet changes in service demand in environments of unpredictability.

Today, as we debate what the next infrastructure component should be, we should fundamentally question whether new infrastructure should be more of what we already have or something that doesn't exist yet. Until we get to that point, we will maintain lock-in and perpetuate systems that we know may already be obsolete.

Notes

[1] L. Megginson, "Lessons from Europe for American Business," *Southwestern Social Science Quarterly* 44, no. 1(1963): 3–13.

[2] D. R. Brennan, P. W. Turnbull, and D. T. Wilson, "Dyadic adaptation in business-to-business markets," *European Journal of Marketing* 37, nos. 11/12 (2003): 1636–1665.

[3] B. S. Chakravarthy, "Adaptation: A promising metaphor for strategic management," *Academy of Management Review* 7, no. 1 (1982): 35–44; L. G. Hrebiniak, and W. F. Joyce, "Organizational Adaptation: Strategic Choice and Environmental Determinism," *Administrative Science Quarterly* 30, no. 3 (1985): 336–349.

[4] D. Garlan, S. W. Cheng, A. C. Huang, B. Schmerl, and P. Steenkiste, "Rainbow: Architecture-Based Self-Adaptation with Reusable Infrastructure," *Computer* 37, no. 10 (2004): 46–54.

[5] B. S. Chakravarthy, "Adaptation: A promising metaphor for strategic management," *Academy of Management Review* 7, no. 1 (1982): 35–44; A.-M. Grisogono, "Success and Failure in Adaptation," in *Sixth International Conference on Complex Systems*, Boston, MA (2006).

[6] Department of Homeland Security (DHS), "Critical Infrastructure Sectors" (2017), retrieved from https://www.dhs.gov/critical-infrastructure-sectors.

[7] American Society of Civil Engineers (ASCE), "Failure To Act: Closing the Infrastructure Investment Gap" (2016).

[8] R. Costanza, R. de Groot, L. Braat, I. Kubiszewski, L. Fioramonti, P. Sutton, S. Farber, and M. Grasso, "Twenty years of ecosystem services: How far have we come and how far do we still need to go?" *Ecosystem Services* 28, Part A (2017): 1–16.

[9] T. Pollard, "Follow the Money: Transportation Investments for Smarter Growth," *Temple Environmental Law & Technology Journal* (2003); D. Shoup, *The High Cost of Free Parking* (New York, NY: American Planning Association, 2011).

[10] T. P. Hughes, *Networks of Power: Electrification in Western Society, 1880-1930* (Baltimore, MD: Johns Hopkins University Press, 1993).

[11] National Research Council, *Privatization of Water Services in the United States* (Washington, DC: National Academies Press, 2002).

[12] G. E. Marchant, B. R. Allenby, and J. R. Herkert, *The Growing Gap Between Emerging Technologies and Legal-Ethical Oversight: The Pacing Problem* (New York, NY: Springer, 2011).

[13] American Society of Civil Engineers (ASCE), "2017 Infrastructure Report Card: A Comprehensive Assessment of America's Infrastructure" (2017).

[14] P. C. D. Milly, J. Betancourt, M. Falkenmark, R. M. Hirsch, Z. W. Kundzewicz, D. P. Lettenmaier, and R. J. Stouffer, "Stationarity Is Dead: Whither Water Management?" *Science* 319, no. 5863 (2008): 573–574.

[15] J. Davis, "Federal Infrastructure Grants: How to Get Back to 'Average?'" *Eno Transportation Weekly* (May 2017).

[16] American Society of Civil Engineers (ASCE), "2017 Infrastructure Report Card: A Comprehensive Assessment of America's Infrastructure" (2017).

[17] J. Menendez, S. Siabil, P. Narciso, and N. Gharaibeh, "Prioritizing Infrastructure Maintenance and Rehabilitation Activities Under Various Budgetary Scenarios," *Transportation Research Record: Journal of the Transportation Research Board* 2361 (2013): 56–62.

[18] C. Shirley, "The Status of the Highway Trust Fund and Options for Paying Highway Spending," testimony before the Committee on Ways and Means, U.S. House of Representatives (June 17, 2015).

[19] D. Paletta, "States Siphon Gas Tax for Other Uses," *The Wall Street Journal* (July 16, 2014).

[20] American Society of Civil Engineers (ASCE), "Failure To Act: Closing the Infrastructure Investment Gap" (2016); American Water Works Association (AWWA), "Buried No Longer: Confronting America's Water Infrastructure Challenge" (2012).

[21] A. Fraser and M. V. Chester, "Environmental and Economic Consequences of Permanent Roadway Infrastructure Commitment: City Road Network Lifecycle Assessment and Los Angeles County" *Journal of Infrastructure Systems* 22, no. 1(2016): 4015018.

[22] J. Garreau, *Edge City: Life on the New Frontier* (New York, NY: Knopf Doubleday, 2011); J. Kornai, "Resource-Constrained Versus Demand-Constrained Systems," *Econometrica* 47, no. 4 (1979): 801.

[23] T. Walski, "A History of Water Distribution," *Journal of the American Water Works Association* 98, no. 3 (2006): 110–121.

[24] S. M. Rinaldi, J. P. Peerenboom, and T. K. Kelly, "Identifying, understanding, and analyzing critical infrastructure interdependencies," *IEEE Control Systems Magazine* 21, no. 6 (2001): 11–25.

[25] L. Chanab, B. El-Darwich, G. Hasbani, and M. Mourad, "Telecom Infrastructure Sharing: Regulatory Enablers and Economic Benefits," Booz Allen Hamilton (2007).

[26] E. G. Amoroso, *Cyber Attacks: Protecting National Infrastructure* (New York, NY: Butterworth-Heinemann, 2013).

[27] D. L. Childers, J. Corman, M. Edwards, and J. J. Elser, "Sustainability Challenges of Phosphorus and Food: Solutions from Closing the Human Phosphorus Cycle," *BioScience* 61, no. 2 (2011): 117–124;P. M. Vitousek, J. D. Aber, R. W. Howarth, G. E. Likens, P. A. Matson, D. W. Schindler, W. H. Schlesinger, and D. G. Tilman, "Human Alteration of the Global Nitrogen Cycle: Sources and Consequences," *Ecological Applications* 7, no. 3

(1997): 737–750; C. J. Vörösmarty and D. Sahagian, "Anthropogenic Disturbance of the Terrestrial Water Cycle," *BioScience* 50, no. 9 (2000): 753.

[28] U.S. National Climate Assessment, *Climate Change Impacts in the United States* (Washington, DC: U.S. Global Climate Change Research Program, 2014).

[29] K. L. Gilroy and R. H. McCuen, "A nonstationary flood frequency analysis method to adjust for future climate change and urbanization," *Journal of Hydrology* 414–415 (2012): 40–48; Y. Tramblay, L. Neppel, J. Carreau, and K. Najib, "Non-stationary frequency analysis of heavy rainfall events in southern France," *Hydrological Sciences Journal* 58, no. 2 (2013): 280–294.

[30] M. Galeotti, "The 'Gerasimov Doctrine' and Russian Non-Linear War," *In Moscow's Shadows* blog (2014); Q. Liang and W. Xiangsui, *Unrestricted Warfare* (Beijing, China: PLA Literature and Arts Publishing House, 1999).

[31] K. Zetter, "Inside the Cunning, Unprecedented Hack of Ukraine's Power Grid," *Wired Magazine* (March 3, 2013).

[32] K. Giles, *Handbook of Russian Information Warfare* (2016), retrieved from http://www.ndc.nato.int/news/news.php?icode=995.

[33] Federal Emergency Management Agency (FEMA), "Hazard Mitigation Planning," (2017), retrieved from https://www.fema.gov/hazard-mitigation-planning.

[34] E. Oughton and P. Tyler, *Infrastructure as a Complex Adaptive System* (Oxford, UK: Oxford University Press, 2013).

[35] S. Arbesman, *Overcomplicated: Technology at the Limits of Comprehension* (New York, NY: Penguin Publishing Group, 2016).

[36] North American Electric Reliability Council, "Technical Analysis of the August 14, 2003, Blackout," Report to the NERC Board of Trustees by the NERC Steering Group (July 13, 2004).

[37] S. Arbesman, *Overcomplicated: Technology at the Limits of Comprehension* (New York, NY: Penguin Publishing Group, 2016).

[38] J. Conklin, *Dialogue Mapping: Building Shared Understanding of Wicked Problems* (West Sussex, UK: Jophn Wiley & Sons, 2006);

P. Willetts, "The Growth in the Number of NGOs in Consultative Status with the Economic and Social Council of the United Nations," (2015), retrieved from http://www.staff.city.ac.uk/p.willetts/NGOS/NGO-GRPH.HTM.

[39] J. Conklin, *Dialogue Mapping: Building Shared Understanding of Wicked Problems* (West Sussex, UK: Jophn Wiley & Sons, 2006); H. Rittel and M. Webber, "Dilemmas in a general theory of planning," *Policy Sciences* 4, no. 2 (1973): 155–169.

[40] B. Allenby, *The Theory and Practice of Sustainable Engineering* (Upper Saddle River, NJ: Pearson Prentice Hall, 2012).

[41] V.-W. Mitchell, "Organizational Risk Perception and Reduction: A Literature Review," *British Journal of Management* 6, no. 2 (1995):115–133; P. Slovic, *The Perception of Risk* (New York, NY: Taylor & Francis, 2016).

[42] E. Hollnagel, J. Pariés, D. D. Woods, and J. Wreathall, eds., *Resilience Engineering in Practice: A Guidebook* (New York, NY: Ashgate, 2011); J. Park, T. P. Seager, P. S. C. Rao, M. Convertino, and I. Linkov, "Integrating Risk and Resilience Approaches to Catastrophe Management in Engineering Systems," *Risk Analysis* 33, no. 3 (2013): 356–367.

[43] A. Hommels, "Studying Obduracy in the City: Toward a Productive Fusion between Technology Studies and Urban Studies," *Science, Technology & Human Values* 30, no. 3 (2005): 323–351.

[44] W. E. Bijker, T. P. Hughes, and T. Pinch, *The Social Sonstruction of Technological Systems: New Directions in the Sociology and History of Technology* (Cambridge, MA: MIT Press, 1987); A. Carse and J. A. Lewis, "Toward a political ecology of infrastructure standards: Or, how to think about ships, waterways, sediment, and communities together," *Environment and Planning A*, 49, no. 1(2017): 9–28.

[45] E. S. Bernardes and M. D. Hanna, "A theoretical review of flexibility, agility and responsiveness in the operations management literature: Toward a conceptual definition of customer responsiveness," *International Journal of Operations & Production Management* 29, no. 1 (2009): 30–53.

[46] S. Meerow, J. P. Newell, and M. Stults, "Defining urban resilience: A review," *Landscape and Urban Planning* 147 (2016): 38–49.

[47] J. Park, T. P. Seager, and P. S. C. Rao, "Lessons in risk-versus resilience-based design and management," *Integrated Environmental Assessment and Management* 7, no. 3 (2011): 396–399.

[48] J. Ahern "From fail-safe to safe-to-fail: Sustainability and resilience in the new urban world," *Landscape and Urban Planning* 100, no. 4 (2011): 341–343; N. Möller and S. O. Hansson, "Principles of engineering safety: Risk and uncertainty reduction," *Reliability Engineering & System Safety* 93, no. 6 (2008): 798–805; J. Park, T. P. Seager, P. S. C. Rao, M. Convertino, and I. Linkov, "Integrating Risk and Resilience Approaches to Catastrophe Management in Engineering Systems," *Risk Analysis* 33, no. 3 (2013): 356–367.

[49] iNEMI, "iNEMI Roadmap," (2017), retrieved from http://www.inemi.org/; S. Pedersen, C. Wilson, G. Pitts, and B. Stotesbery, "Electronics Industry Environmental Roadmap," in *Proceedings of the 1995 IEEE International Symposium on Electronics and the Environment ISEE* (1996).

[50] S. Pedersen, C. Wilson, G. Pitts, and B. Stotesbery, "Electronics Industry Environmental Roadmap," in *Proceedings of the 1995 IEEE International Symposium on Electronics and the Environment ISEE* (1996).

[51] H. Bulkeley, V. Castán Broto, and G. A. S. Edwards, *An Urban Politics of Climate Change: Experimentation and the Governing of Socio-Technical Transitions* (New York, NY: Routledge, 2014); A. Hommels, "Studying Obduracy in the City: Toward a Productive Fusion between Technology Studies and Urban Studies," *Science, Technology & Human Values* 30, no. 3 (2005): 323–351.

[52] A. C. Lemer, "Infrastructure Obsolescence and Design Service Life," *Journal of Infrastructure Systems* 2, no. 4 (1996): 153–161.

[53] S. Lohr, "G.E., the 124-Year-Old Software Start-Up," *New York Times* (Sept. 8, 2016).

[54] J. Lynch and K. Loh, "A Summary Review of Wireless Sensors and Sensor Networks for Structural Health Monitoring," *The Shock and Vibration Digest* 38 (2006): 91–128.

[55] M. Frankowiak, R. Grosvenor, and P. Prickett, "A review of the evolution of microcontroller-based machine and process monitoring," *International Journal of Machine Tools and Manufacture* 45, nos. 4–5 (2005): 573–582.

[56] J. Resnik, "Virtual assembly lines are making the auto industry more flexible," *Ars Technica* (Sept. 6, 2016).

[57] T. Neuberger and S. Weston, "Variable frequency drives: energy savings for pumping applications," Industry Application IA04008002en, Eaton Industries (2012); D. Roethemeyer and D. Yankaskas, "Evolution of Motor and Variable Frequency Drive Technology," in *Summer Study on Energy Efficiency in Industry Conference Proceedings* (1995), 541–552.

[58] R. Kenny, "Replacing In-Pavement Loops with Video Detection," *International Municipal Signal Association* (May/June 2004): 32–35.

[59] J. Gonder, M. Earleywine, and W. Sparks, "Analyzing Vehicle Fuel Saving Opportunities through Intelligent Driver Feedback," *SAE International Journal of Passenger Cars* 5, no. 2 (2012): 450–461.

[60] S. Kaplan and B. J. Garrick, "On The Quantitative Definition of Risk," *Risk Analysis* 1, no. 1 (1981): 11–27.

[61] J. Park, T. P. Seager, and P. S. C. Rao, "Lessons in risk-versus resilience-based design and management," *Integrated Environmental Assessment and Management* 7, no. 3 (2011): 396–399.

[62] S. Chung, R. K. Rainer, and B. Lewis, "The Impact of Information Technology Infrastructure Flexibility on Strategic Alignment and Application Implementations," *The Communications of the Association for Information Systems* 11, no. 1 (2003): 191–206.

[63] N. B. Duncan, "Capturing Flexibility of Information Technology Infrastructure: A Study of Resource Characteristics and Their Measure," *Journal of Management Information Systems* 12 no. 2 (1995): 37–57.

[64] N. B. Duncan, "Capturing Flexibility of Information Technology Infrastructure: A Study of Resource Characteristics and Their Measure," *Journal of Management Information Systems* 12 no. 2 (1995): 37–57.

[65] C. Y. Baldwin and K. B. Clark, "Modularity in the Design of Complex Engineering Systems," in *Complex Engineered Systems: Science Meets Technology*, D. Braha, A. A. Minai, and Y. Bar-Yam, eds. (Berlin, Germany: Springer Berlin Heidelberg, 2006).

[66] N. B. Duncan, "Capturing Flexibility of Information Technology Infrastructure: A Study of Resource Characteristics and Their Measure," *Journal of Management Information Systems* 12 no. 2 (1995): 37–57.

[67] C. Y. Baldwin and K. B. Clark, "Modularity in the Design of Complex Engineering Systems," in *Complex Engineered Systems: Science Meets Technology*, D. Braha, A. A. Minai, and Y. Bar-Yam, eds. (Berlin, Germany: Springer Berlin Heidelberg, 2006).

[68] C. Freeman and F. Louçã, *As Time Goes By: From the Industrial Revolutions to the Information Revolution* (Oxford, UK: Oxford University Press, 2001).

[69] M. J. Hatch, *Organization Theory: Modern, Symbolic, and Postmodern Perspectives* (Oxford, UK: Oxford University Press, 1997); R. P. Vecchio, *Organizational Behavior: Core Concepts* (United Kingdom: Thomson/South-Western 2006).

[70] L. Donaldson, *The Contingency Theory of Organizations* (Thousand Oaks, CA: Sage Publications, 2001); B. Sherehiy, W. Karwowski, and J. K. Layer, "A review of enterprise agility: Concepts, frameworks, and attributes," *International Journal of Industrial Ergonomics* 37, no. 5 (2007): 445–460.

[71] B. Sherehiy, W. Karwowski, and J. K. Layer, "A review of enterprise agility: Concepts, frameworks, and attributes," *International Journal of Industrial Ergonomics* 37, no. 5 (2007): 445–460.

[72] B. Sherehiy, W. Karwowski, and J. K. Layer, "A review of enterprise agility: Concepts, frameworks, and attributes," *International Journal of Industrial Ergonomics* 37, no. 5 (2007): 445–460.

[73] W. Ashby, *Design for a Brain: The Origin of Adaptive Behaviour* (Chapman & Hall and Science Paperbacks, 1960).

[74] P. R. Lawrence and J. W. Lorsch, *Organization and Environment: Managing Differentiation and Integration* (Cambridge, MA: Harvard Business School Press, 1967).

[75] K. E. Weick and R. E. Quinn, "Organizational Change and Development," *Annual Review of Psychology* 50, no. 1 (1999): 361-386.

[76] A. Dastmalchian, "The concept of organizational flexibility: exploring new direction," in *Organizational Flexibility, Proceedings of a Colloquium* (Victoria, British Columbia: University of Victoria, 1993).

[77] A. Dastmalchian and P. Blyton, "Organizational flexibility in cross-national perspective: an introduction," *The International Journal of Human Resource Management* 9, no. 3 (1998): 437–444; Kalleberg, 2001; B. Sherehiy, W. Karwowski, and J. K. Layer, "A review of enterprise agility: Concepts, frameworks, and attributes," *International Journal of Industrial Ergonomics* 37, no. 5 (2007): 445–460.

[78] B. Sherehiy, W. Karwowski, and J. K. Layer, "A review of enterprise agility: Concepts, frameworks, and attributes," *International Journal of Industrial Ergonomics* 37, no. 5 (2007): 445-460.

[79] M. A. Delucchi and M. Z. Jacobson, "Providing all global energy with wind, water, and solar power, Part II: Reliability, system and transmission costs, and policies," *Energy Policy* 39, no. 3 (2011): 1170–1190; M. Z. Jacobson and M. A. Delucchi, "Providing all global energy with wind, water, and solar power, Part I: Technologies, energy resources, quantities and areas of infrastructure, and materials," *Energy Policy* 39, no. 3 (2011): 1154–1169; D. McCollum and C. Yang, "Achieving deep reductions in U.S. transport greenhouse gas emissions: Scenario analysis and policy implications," *Energy Policy* 37, no. 12 (2009): 5580–5596.

[80] American Society of Civil Engineers (ASCE), "Failure To Act: Closing the Infrastructure Investment Gap" (2016).

[81] C. Wolmar, *The Great Railroad Revolution: The History of Trains in America* (New York, NY: PublicAffairs, 2012).

[82] M. J. Nahlik and M. V. Chester, "Policy Making Should Consider Time-Dependent Greenhouse Gas Benefits of Transit-Oriented Smart Growth," *Transportation Research Record: Journal of the Transportation Research Board* 2502 (2015): 53–61.

5

WHAT IS CLIMATE-READY INFRASTRUCTURE?*

Mikhail Chester, Braden Allenby, and Samuel Markolf

The most recent international report on climate change paints a picture of disruption to society unless there are drastic and rapid cuts in greenhouse gas emissions.

Although it's early days, some cities and municipalities are starting to recognize that past conditions can no longer serve as reasonable proxies for the future. This is particularly true for the country's infrastructure. Highways, water treatment facilities and the power grid are at increasing risk to extreme weather events and other effects of a changing climate.

The problem is that most infrastructure projects, including the Trump administration's infrastructure revitalization plan, typically ignore the risks of climate change. In our work researching sustainability and infrastructure, we

* Republished from M. Chester, B. Allenby, and S. Markolf, "What is climate-ready infrastructure? Some cities are starting to adapt," *The Conversation* (Oct. 22, 2018).

encourage and are starting to shift toward designing man-made infrastructure systems with adaptability in mind.

Designing for the Past

Infrastructure systems are the front line of defense against flooding, heat, wildfires, hurricanes and other disasters. City planners and citizens often assume that what is built today will continue to function in the face of these hazards, allowing services to continue and to protect us as they have done so in the past. But these systems are designed based on histories of extreme events.

Pumps, for example, are sized based on historical precipitation events. Transmission lines are designed within limits of how much power they can move while maintaining safe operating conditions relative to air temperatures. Bridges are designed to be able to withstand certain flow rates in the rivers they cross. Infrastructure and the environment are intimately connected.

Now, however, the country is more frequently exceeding these historical conditions and is expected to see more frequent and intense extreme weather events. Said another way, because of climate change, natural systems are now changing faster than infrastructure.

How can infrastructure systems adapt? First let's consider the reasons infrastructure systems fail at extremes:

- The hazard exceeds design tolerances. This was the case of Interstate 10 flooding in Phoenix in fall 2014, where the intensity of the rainfall exceeded design conditions.
- During these times there is less extra capacity across the system: When something goes wrong there are fewer options for managing the stressor, such as rerouting flows, whether it's water, electricity or even traffic.

- We often demand the most from our infrastructure during extreme events, pushing systems at a time when there is little extra capacity.

Gradual change also presents serious problems, partly because there is no distinguishing event that spurs a call to action. This type of situation can be especially troublesome in the context of maintenance backlogs and budget shortfalls which currently plague many infrastructure systems. Will cities and towns be lulled into complacency only to find that their long-lifetime infrastructure systems are no longer operating like they should?

Currently the default seems to be securing funding to build more of what we've had for the past century. But infrastructure managers should take a step back and ask what our infrastructure systems need to do for us into the future.

Agile and Flexible by Design

Fundamentally new approaches are needed to meet the challenges not only of a changing climate, but also of disruptive technologies.

These include increasing integration of information and communication technologies, which raises the risk of cyberattacks. Other emerging technologies include autonomous vehicles and drones as well as intermittent renewable energy and battery storage in the place of conventional power systems. Also, digitally connected technologies fundamentally alter individuals' cognition of the world around us. Consider how our mobile devices can now reroute us in ways that we don't fully understand based on our own travel behavior and traffic across a region.

Yet our current infrastructure design paradigms emphasize large centralized systems intended to last for decades and that can withstand environmental hazards to a preselected level of risk. The problem is that the level of risk is now uncertain because the climate is changing, sometimes in ways that are not very well-understood. As such, extreme events forecasts may be a little or a lot worse.

Given this uncertainty, agility and flexibility should be central to our infrastructure design. In our research, we've seen how a number of cities have adopted principles to advance these goals already, and the benefits they provide.

In Kuala Lampur, traffic tunnels are able to transition to stormwater management during intense precipitation events, an example of multifunctionality.

Across the United States, citizen-based smartphone technologies are beginning to provide real-time insights. For instance, the CrowdHydrology project uses flooding data submitted by citizens that the limited conventional sensors cannot collect.

Infrastructure designers and managers in a number of U.S. locations, including New York, Portland, Miami and southeast Florida, and Chicago, are now required to plan for this uncertain future—a process called roadmapping. For example, Miami has developed a $500 million plan to upgrade infrastructure, including installing new pumping capacity and raising roads to protect at-risk oceanfront property.

These competencies align with resilience-based thinking and move the country away from our default approaches of simply building bigger, stronger or more redundant.

Planning for Uncertainty

Because there is now more uncertainty with regard to hazards, resilience instead of risk should be central to infrastructure design and operation in the future. Resilience means systems can withstand extreme weather events and come back into operation quickly.

This means infrastructure planners cannot simply change their design parameter—for example, building to withstand a 1,000-year event instead of a 100-year event. Even if we could accurately predict what these new risk levels should be for the coming century, is it technically, financially or politically feasible to build these more robust systems?

This is why resilience-based approaches are needed that emphasize the capacity to adapt. Conventional approaches emphasize robustness, such as building a levee that is able to withstand a certain amount of sea level rise. These approaches are necessary but given the uncertainty in risk we need other strategies in our arsenal.

For example, providing infrastructure services through alternative means when our primary infrastructure fail, such as deploying microgrids ahead of hurricanes. Or, planners can design infrastructure systems such that when they fail, the consequences to human life and the economy are minimized.

This is a practice recently implemented in the Netherlands, where the Rhine delta rivers are allowed to flood but people are not allowed to live in the flood plain and farmers are compensated when their crops are lost.

Uncertainty is the new normal, and reliability hinges on positioning infrastructure to operate in and adapt to this uncertainty. If the country continues to commit to building last century's infrastructure, we can continue to expect failures of these critical systems and the losses that come along with them.

6

INFRASTRUCTURE AND THE ENVIRONMENT IN THE ANTHROPOCENE*

Mikhail Chester, Samuel Markolf, and Braden Allenby

Infrastructure has in many ways reflected human social and demographic structures. Infrastructures have for much of human history been the designed and built set of technological systems that mediate between humans, their communities, and their broader environment. Early hunter-gatherer humans had minimal-to-no shared infrastructure. The Neolithic revolution brought agriculture and its associated technologies as infrastructure, as well as town systems. During the classical era infrastructure became quite sophisticated but mostly remained within empires. An exception being transportation, where marine fleets and caravans (e.g., Roman, Islamic, Chinese empires) created trade infrastructure that extended far beyond polit-

* Adapted from M. V. Chester, S. Markolf, and B. Allenby, "Infrastructure and the environment in the Anthropocene," *Journal of Industrial Ecology* 23, no. 5 (2019): 1006–1015, with permission from Wiley.

ical boundaries. The Romans brought advancements in hydraulics (including aqueducts), pavements (for roads), communications (mail), and energy use (petroleum), technologies that would not see widespread use again for centuries after the empire's collapse.[1] Agricultural technologies including irrigation (e.g., dikes and dams) and farming tools were developed throughout this period by Chinese dynasties. The modern globalized era and now the Anthropocene has accelerated the pace and increased the scale of technology, where the distinctions between nature and infrastructure are blurring.

The infrastructures that evolved largely since the Industrial Revolution were viewed as separate systems from these larger natural systems. This was especially true in cities where a large number of people and infrastructure coalesced, and there was an increasing need to keep environmental hazards at bay while extracting large quantities of resources from the environment. The interfaces of infrastructure and the environment were largely defined as designing 1) with resources from the environment; 2) to remove and transform resources from the environment; and 3) to withstand environmental perturbations. These interfaces largely determined (and continue to determine) the location and design of infrastructure. Unexpected changes in natural systems make slow-changing (i.e., obdurate) infrastructure obsolete, as evidenced by a third intake pipe at Lake Mead completed in 2015 for Las Vegas—the first two becoming obsolete as water levels have dropped (we acknowledge that there are deep and unique complexities surrounding water policy in the Western United States, like for most large infrastructure).

Furthermore, infrastructure are at risk to changing environmental hazards such as climate change.[2] Symptoms of the problem are that 1) the resources that we use to construct infrastructure are becoming more and more global (and less local), escalating their costs and reducing ease of access; 2) the infrastructure that move resources are largely inflexible and any changes in resource supply and/or external conditions means there is a mismatch in utilization (and we often have difficulty predicting how much we'll use and by when); 3) the cycle time of changes in human, built, and natural systems has accelerated, resulting in legacy infrastructure which is still in place but increasingly obsolete, if not dysfunctional; and 4) climate and other earth systems changes threaten the model that we use that infrastructure must be able to continue to provide reliable service despite perturbations. The crux of the problem is not change itself but rather the volatility, scale, and cycle time associated with change.

The scale and scope of human activities has changed dramatically in the past century and show signs of accelerating into the future—particularly with regard to trends of increasing urbanization.[3] Economic development and other aspects of global history were certainly very different around the world (e.g., Europe, the United States, and Asia), but the current convergence reflects the competitive global environment that we see today.

Prior to the Anthropocene, human activities and their impacts were small relative to earth systems. Separating infrastructure and the environment made sense since human caused environmental impacts were within the capacity of the environment to absorb quickly. As human society has progressed, the significance of our role as hyperkeystone species has emerged.[4] Keystone species exhibit disproportionate and often unexpected effects on natural systems, and are found in every major habitat. The term hyperkeystone has been coined to describe the influence that humans have over not just other keystone species but also natural systems.[5] Across many earth systems—climate, water, nitrogen, phosphorous, precious metals, etc.—evidence accumulates that human activities are creating global and unpredictable impacts that create unintended consequences.[6] The reach of human activities has grown so large that infrastructure, and the activities and technologies that they support, are directly and at scale managing natural systems.

Examples of infrastructure and the environment converging are found across regional and global scales. Consider urban water supply. Arizona is a state with diverse climates and geography but whose population largely exists in semi-arid desert regions of the state. Water supply to this population, contrary to popular belief, is diverse with 17% from in-state rivers and associated reservoirs, 40% from groundwater, 3% reclaimed, and 40% imported from the Colorado River.[7] A large portion of imported water is banked. Wastewater is sent to either power plants of used for recharging groundwater aquifers.

Similarly, land use laws and regulations designed to protect New York City's drinking water sources have far-reaching implications elsewhere in the state.[8] In other words, New York's nature has been designed as part of water supply infrastructure. Water in California is also tightly designed: the mountains store snow, which becomes runoff, which is heavily managed as it moves toward the ocean to optimize its use by people.

Like many other regions, natural flows from precipitation are now highly managed and non-stationarity—i.e., historical flows are becoming less and less relevant for predicting future flows—is the norm. Watersheds—particularly for urban water supply—are intensely managed. Thinking of the environment as water infrastructure is necessary for maintaining supply. In many ways, the dichotomy between infrastructure and the "environment" no longer exists; they are one and the same.

Wildlife in some contexts are becoming entirely managed by humans through technology. The upstream river migration of salmon in the Northwest United States is controlled by management techniques including ladders, a need that emerged after dams were constructed.[9] Aquaculture (ocean fish farming) represents the development of marine infrastructure and associated technologies to directly manage populations.[10]

Anthropogenic carbon dioxide, driven increasingly by urban activity,[11] has resulted in the atmosphere becoming a design space for technology. Efforts to manage atmospheric carbon dioxide concentrations involve the deployment of new transportation and energy technologies to reduce additions of greenhouse gases, as well as technologies to remove carbon dioxide (i.e., carbon capture and storage). There are even discussions on reducing incoming solar radiation to manage the energy balance of the planet.[12] When it comes to managing the energy balance of the planet, infrastructure and the atmosphere are becoming one. The atmosphere is becoming testbed of emergent and planned human activities. Going forward we may not be able to afford the inefficiency (from a human perspective) and unexpected transients that such a sloppy approach to infrastructure implies.

Fundamentally, the models that we have used to build and operate infrastructure have separated infrastructure and the environment as independent systems. This framing was a useful simplification when natural systems were not affected in significant ways by human activity. However, as human activities have grown, their reach into the environment has increased to the point where some natural systems are now largely managed by humans. And as human activities continue to grow and the need to secure reliable natural resources and services increases, it is likely that we'll see increased management of environmental systems. As such, in the long term, major earth systems will continue to become part of our infrastructure designs. In the Anthropocene, infrastructure will become how we design nature and the planet, integrating relevant human, natural, and built systems in design through technologies that support local, regional, and even global activities.

Infrastructure Obduracy

Our core infrastructure systems have developed a relatively high amount of obduracy over time. Broadly speaking, obduracy refers to rigidity and difficulty or resistance to change.[13] Perhaps nowhere is obduracy more prevalent or impactful than in cities. "Despite the fact that cities are considered to be dynamic and flexible spaces...it is very difficult to radically alter a city's design: once in place, urban structures become fixed, obdurate."[14] Obduracy arises for several reasons: 1) changes in demand for the services provided by our infrastructure systems have historically been relatively gradual and predictable (we demand water, power, and vehicle travel much like we did decades, even a century ago); 2) from a financial perspective, the high upfront capital cost associated with many infrastructure projects often necessitates a longer lifespan; 3) infrastructure systems often operate in economic settings with minimal competition (and impetus for innovation)—many infrastructure systems are either publicly funded or natural monopolies; 4) many infrastructure systems are installed under the assumption of stable environmental and societal conditions (i.e., stationarity is assumed); and 5) many infrastructure systems are constructed of long-lasting materials and legal structures (such as right-of-ways) that have inherently long cycle times.

Historically, obdurate infrastructure has provided some level of benefit in terms of reliability, predictability, safety, and efficiency. But, as seen in the example of waste incineration in Goteborg, Sweden, the obduracy that contributed to and emerged from once desirable services (diverting waste from landfills and using it as a fuel source), can eventually inhibit the adoption of innovative and more preferable solutions (e.g., recycling, waste reduction programs, generating electricity from wind or solar power, etc.).[15] In the context of the Anthropocene, obduracy may become problematic for a couple different reasons. First, it appears to be a major factor in the growing and complex relationship that infrastructure have with earth and social systems. For example, cities such as Detroit that have experienced large population decreases have had to grapple with infrastructure that is now underutilized and poorly funded. The inability of the infrastructure (and related institutions and practices) to adjust and adapt to the decrease in population and demand has created challenges related to funding, maintenance, consumer affordability, public health, and environmental degradation.[16] The obdurate nature of most infrastructure (and related institutions) can also conflict with sustainability oriented goals such as increased urban density, transit oriented development, and distributed solar power. Similarly, the obduracy of many infrastructure systems contributes to the ongoing expansion of resource extraction efforts.[17] For example, the fact that the transportation sector has been consistently and nearly exclusively powered by fossil fuels for over a century has contributed to the need for increasing efforts to secure fuel (e.g., off shore oil drilling, hydraulic fracturing, plant-based ethanol, etc.). Finally, the longer certain infrastructure systems are around, the more opportunity there is for social, cultural, and other technological systems to couple and co-evolve with the infrastructure systems—thereby increasing complexity and scale.[18]

The other concern with infrastructure obduracy in the context of the Anthropocene is that the cycles of infrastructure evolution (i.e., maintenance, renovation, and replacement schedules) may increasingly become out of sync with environmental (e.g., climate change, resource accessibility and availability) and cultural (e.g., societal demand, preferences, and needs) cycle times. Historically, infrastructure have had cycle times that align well with our social, economic, and technological systems. However, social, environmental, and technological changes appear to be accelerating at a faster rate than infrastructure systems.[19] For example, the growing trend toward electric vehicles[20] means that the economic, politic, cultural, and environmental viability of a project like the Keystone XL pipeline may come into question over the coming decades. Similarly, with an increasing push toward online shopping, ride-share systems, and autonomous vehicles, how viable is it to continue to size much of our parking infrastructure—and all of the land-use and environmental issues associated with it[21]—to handle high volumes that infrequently occur (e.g., the largest shopping days of the year, special events, etc.)? Aside from accelerating technological and social changes, fluctuations in nitrogen, phosphorus, water, carbon, climate, and natural resource cycles should increasingly challenge the assumption that infrastructure can be designed, installed, and operated under stable environmental conditions. For example, in response to sea level rise and tidal flooding events, the City of Miami Beach has spent roughly $100 million (as part of a larger $500 million effort) to raise street levels and install pumping stations throughout the island. At the same time, concerns have been raised about the impact that the effluent from the pumping stations is having on water quality and ecosystem health in Biscayne Bay.[22] The issue of non-stationarity—the concept that historical conditions are increasingly a poor predictor of future conditions[23]—raises important questions about the design, implementation, and obduracy of our infrastructure systems. Engineers, over the past century, have generally assumed stationarity in environmental systems in order to design for robustness,[24] an assumption that was non-problematic so long as the life time of infrastructure components and the larger environmental systems was roughly equivalent. The stationarity assumption was reasonable because the cycle times of infrastructure (i.e., how quickly we change infrastructure and its associated technologies) were roughly in sync with the environment. However, as environmental (and other) conditions begin to change more rapidly (e.g., climate change; "dead zones" in the Gulf of Mexico), such implicit assumptions no longer hold. Moving to a system of designing for future conditions that are difficult to anticipate seems to be in order. In doing so, decisions will need to be made about what infrastructure life span and future conditions should be assumed, especially given the uncertainty in financial and environmental conditions.[25] This challenge is not specific to climate but also other earth systems that have become profoundly affected by humans.

Ultimately, one of our primary challenges will be meshing infrastructure cycle times (through concepts like flexibility, agility, and adaptability) with the much more rapid cycle times of coupled social, economic, and environmental systems. Similarly, it will increasingly become a challenge for infrastructure managers and practitioners to grapple with the fact that infrastructure decisions are having a growing influence on earth systems. And as urban populations grow, cities will increasingly drive changes in the environment. For example, it has been shown that a relatively modest commitment to ethanol fuel in the transportation sector can have significant implications on global agricultural practices, land use, water and nutrient cycles, and the carbon cycle.[26] Similarly, urban areas typically have higher market shares of electric vehicles (EVs) than the national average.[27] To the extent that the EV market continues to grow as expected and cities remain at the forefront of this growth, urban areas will likely be at the center of major alterations to the material flows and environmental impacts of the electricity sector, the petro-chemical sector, and global supply chains of earth elements like lithium, cobalt, and nickel (i.e., crucial components of batteries). This is not to say that cities are necessarily the cause of negative environmental outcomes, but instead should be viewed as a major driver of the consumption of the policies and complexities that have been introduced in the city and more broadly. Such systemic implications, which we know are there and will be manifested even if we cannot predict specifics, must be part of infrastructure design and management in the Anthropocene. There was a time when ignoring systems-level implications (e.g., the impact of U.S. ethanol use and policy on global food prices, land use patterns, and environmental quality) might have been appropriate to simplify analysis and management. However, that appears to no longer be the case. A major component of addressing these challenges will be overcoming lock-in – the idea that today's systems are constrained by past decisions, even if the context of the past choices is no longer relevant or if new alternatives have emerged that may be more effective.[28] In doing so, it will be important to recognize that changes will not only need to be made to the physical configuration of infrastructure, but also to the institutional, financial, and cultural forces associated with the physical infrastructure. It will also be important to attempt to reconcile with the growing complexity that arises from obdurate and interconnected infrastructure and social systems operating at increasingly large scales.

Infrastructure in the Anthropocene

Our conceptual models of how infrastructure interface with the environment will need to be shifted and we will need to move toward integrated management of technological and ecological systems in cities. We are already seeing, and can expect acceleration of, natural systems moving toward being increasingly human managed. This will require new kinds of engineering skills and professional niches, not to mention engineering education. These managed resources will increasingly fall under the purview of engineering. Engineering, which has traditionally treated the environment as a separate entity that either provides resources or creates hazards, will increasingly need the competencies to design and manage techno-environment (and even social) systems as one. Examples of this are already apparent. Apple hires engineers to ensure that they can design and manufacture given the availability of rare earth metals.[29] Cargill, a major global food producer, has invested in cultured meat.[30]

The coming century will be defined by how we view and position infrastructure in the Anthropocene. As we increasingly manage natural systems, our infrastructure become our environment—particularly in urban settings. With this transition different principles will emerge. Recognizing the complexity (defined partly by an inability to predict emergent behaviors as a result of feedbacks and nonlinearities) of techno-environment systems, emphases will need to shift away from optimizing and toward satisficing.[31] If we allow existing models of infrastructure to persist and a perpetuation of thinking that separates infrastructure from the environment then we can expect 1) cycle times of infrastructure that don't match the environment and our changing needs/technologies; and 2) greater failures (and costs) of infrastructure. Agile (physical structure and the rules, policies, norms, and actors who manage and operate it, will need to be able to maintain function in a non-stationarity future) and flexible (ability to meet changing demands (in the face of both predictable and unpredictable challenges) infrastructure will be key to adapting infrastructure in the short term.[32]

Thomas Kuhn in his classic book *The Structure of Scientific Revolutions*[33] made an important distinction between periods of "normal science," where incremental advances are made within accepted frameworks of theory and belief, and "paradigm shifts," where the frameworks themselves are overthrown, and entirely new ones introduced. Examples in science include the shift from Newtonian physics to quantum mechanics, and, in geology, the shift to plate tectonics from a belief in a static and fixed planetary structure. What we are asserting here is that engineering at the dawn of the Anthropocene must undergo a paradigm shift, in the Kuhnian sense: it is no longer adequate to view infrastructure as simply localized artifacts within an unchanging exogenous human, natural, and built context. Rather, engineering, and infrastructure, are now means by which that context is itself designed and shaped.

Design for adaptive capacity. Infrastructure that is able to quickly respond to changing environmental conditions in the future will be much more adept at adapting to changing demands.[34] The paradigms that define our current infrastructure, and the inherent obduracy that is created, are an artifact of the traditional paradigms that separate infrastructure from the environment. These existing paradigms are insufficient to address the growing complexities of our world; our infrastructure so far hasn't needed to be agile and flexible. But in the future, with rapid and unpredictable changes in earth systems, accelerated disruptive technologies, and the potential for conflict over increasingly scarce local resources, infrastructure will need to be reimagined. For example, the SMART Tunnel in Kuala Lumpur, Malaysia serves the dual purposes of improving mobility and diverting floodwaters away from city center during storms. Under normal conditions, the tunnel serves as a motorway. Under moderate storm conditions, half of the tunnel diverts floodwater while the other half remains operational to motorists. Under extreme storm conditions, automated systems activate water-tight gates and allow to whole tunnel to divert and temporarily storm floodwater.[35] Non-stationarity in many domains can be expected to define the Anthropocene. The competencies needed to produce agile and flexible infrastructure may include multifunctionality, roadmapping, focus on software over hardware, resilience-based thinking, compatibility, connectivity, and modularity of components, organic and change-oriented management, and transdisciplinary education.[36] These competencies exist in industries such as ICT and manufacturing which have shown an ability to quickly adapt to changing demands. Given the obdurate nature of infrastructure, the capacity to adapt (i.e., resilience) becomes critical for addressing some of the complexity of our changing environments, including earth systems such as climate change, and resource availability. Agile and flexible infrastructure might decrease the cycle times of infrastructure so they are on par or ahead of changes in the environment. They would allow adjustment in real time to unpredictable change, thus maintaining more efficient system integration and operation.

Design for complexity. While adaptation strategies can address environmental perturbations, the paradigms that we use to design and operate infrastructure will need to shift toward working with complex techno-environmental systems. Infrastructure by themselves are becoming more and more complex as they grow to encompass old and new technologies, become interdependent, and replace hardware with software. This is particularly true in cities where large numbers of technical and social infrastructure come together, often with decades or centuries old systems. Understanding how a small perturbation cascades through and across infrastructure is becoming increasingly challenging. But complexity is not just the result of hardware, but of interconnected institutional and environmental systems too. As is the case of Arizona, designing water conveyance and distribution systems for urban water delivery is an exercise of hydrological management with direct implications for economic growth and food production. If we choose to deploy atmospheric carbon capture and storage technologies, we are explicitly designing the carbon dioxide content of the atmosphere, and implicitly designing atmospheric dynamics, climate patterns, and ocean circulation patterns. And we may not fully grasp the dynamics of these systems and resulting effects, which may impact other infrastructure and people. This reinforces the need for flexibility and agility, which are required to respond to unpredictable shifts in system state. Emerging technologies, such as artificial intelligence, may be increasingly available to help engineers deal with such complexity rapidly and effectively. However, few engineers are currently trained in using these tools effectively.

Complexity and uncertainty will need to be central to the training of those who design and operate infrastructure in the Anthropocene. Infrastructures encompass a rich diversity of components, organizations, and rules. They include the layering of technologies over decades, some of which despite their antiquated capabilities continue to be used today. For example, despite some difficulties associated with finding replacement parts, equipment dating back to the 1930s enables the reliable signaling and control in the New York City subway.[37] The confluence of these factors means that "we often are left with only the extremes of understanding: either a general notion of how the thing works, even if its innards are at best murky to us, or an examination of its bits and pieces, without an inkling of how it all fits together and how we can expect it to behave."[38]

Adding to this complexity is the non-stationarity in future environmental and other factors. Yet those who design and operate infrastructure are often taught based on stationary and reductionist approaches.[39] Going forward, these infrastructure managers will need to accept that they cannot fully predict the emergent behavior of perturbed systems, or how the systems will be utilized or needed in the future. Unpredictability will be the new normal. As such, competencies around design and management under deep uncertainty will be essential. Resilience-based thinking that emphasizes adaptive capacity, or the ability to move between different approaches for design and management, should be central to academic and job training.

Changing infrastructure takes time and deploying new paradigms will require us to address forces that lock in current practices. For many resources, the means to deliver adequate supply is technologically feasible but implementation is slowed by non-technical forces. For example, the technologies to deliver potable water over long distances are established (e.g., long-distance conveyance and desalination). But bringing them online on time as part of future water supply requires addressing planning, permitting, financing, rights, and other challenges. If not done properly or expeditiously, the transition period between technologies could be costly and dysfunctional.

The difficulty of change in complex adaptive systems can be viewed through dissipative structure theory. Complex adaptive systems are self-organizing, they drop less valuable elements (dissipation) to incorporate new elements, thereby contributing to the conditions for adapting and growing.[40] We can apply the dissipative structure theory to infrastructure growth and change. To change infrastructure on large scales quickly means that we in many ways need to work against the current momentum designed around technologies that consume abundant and finite resources. We need to accelerate the dissipation of elements that are reliant on a consumption model that focuses on these resources in what we anticipate will be a new model, while at the same time recognizing the complexity, that biofeedstocks and carbon capture and storage (for example) may reverse our views on particular technologies. While humans may have the technical means to institute these changes, the forces that prevent this dissipation from happening (financial, cultural, political, etc.) have high inertial resistance to change, and thus allow the current system to continue.

To design, manage, and work within complex techno-environmental systems, engineering domains will need broad competencies and a recognition that complexity is likely to increase. As we accept that infrastructure and the environment have in some ways and are in many ways becoming one in the same, it will be necessary for engineers to not only recognize this but to work within it. In addition to the usual disciplinary expertise, expertise around material flows, earth systems, impact analysis, and others will be needed to create a systemic perspective of technologies and identify the transition, scarcity, and other challenges that may affect the systems.

Reframing Infrastructure

Infrastructure, particularly in engineering, has been synonymous with the physical implementation of structures that provide services to society, a definition that carries less and less meaning as technology progresses and we make new progress toward intentionally designing systems that in the past were not steered by humans. As this intentional design continues and even accelerates in the Anthropocene, our definition of infrastructure will need to change. Cyberspace, technological capability, health care, and other systems that we attempt to increasingly manage and interconnect with other systems will become design spaces. While our current gray infrastructure may indeed persist, albeit in possibly very different forms, it may exist as part of complex systems that provide services in very different ways than what we've experienced in the past.

Currently, engineers are taught to design in response to a set of objectives and constraints which, even in complex domains, can be quantified. But what happens when we know we're doing design that is intended to extend well into a future that is fundamentally unpredictable, and where the design objectives and constraints we have now are at best contingent? How do we design infrastructure that is in some meaningful sense contingent on future conditions we might not even be able to conceive (e.g., urban infrastructure for the AI city? Or, less speculatively, urban transportation infrastructure, designed for decades, as we shift to autonomous vehicles providing a service rather than owned vehicles parked for 95% of the time?).

The management of infrastructure in the Anthropocene and the societal goals it enables is not a challenge for the engineering or infrastructure professions exclusively. Society as a whole must recognize that the scale and scope of human activities are becoming so large, and often locked-in, that fast and large-scale change is becoming difficult, if not impossible. Designing for adaptive capacity and complexity is not simply a job for engineers. Collective action and broad changes in worldviews are needed, tasks that have proven difficult for other professions, such as climate scientists. The emerging field of cultural evolution[41] suggests that worldviews co-evolve with the societies they support and that, absent catastrophic collapse, they are hard to change.

As the tendrils of human activity reach further and our natural systems become increasingly managed systems, casting infrastructure as a coupled human-environmental system will become increasingly important. No longer should we view infrastructure as separate from the environment; they are becoming inexorably linked in ways that will require us to manage them as single (and complex) systems.

Notes

[1] T. Walski, "A History of Water Distribution," *Journal of the American Water Works Association* 98, no. 3 (2006): 110–121.

[2] J. M. Melillo, T. C. Richmond, and G. W. Yohe, eds., *Climate Change Impacts in the United States: The Third National Climate Assessment* (Washington, DC: U.S. Global Climate Change Research Program, 2014).

[3] UNESCO, "Global Trend Towards Urbanisation" (2010); P. Romero-Lankao and D. Dodman, "Cities in transition: Transforming urban centers from hotbeds of GHG emissions and vulnerability to seedbeds of sustainability and resilience," *Current Opinion in Environmental Sustainability* 3, no. 3 (2011): 113–120; T. Elmqvist, M. Fragkias, J. Goodness, B. Güneralp, P. J. Marcotullio, R. I. McDonald, S. Parnell, M. Schewenius, M. Sendstad, K. C. Seto, and C. Wilkinson, eds., *Urbanization, Biodiversity and Ecosystem Services: Challenges and Opportunities* (New York, NY: Springer, 2013); UN Department of Economic and Social Affairs, "World Urbanization Prospects" (2014).

[4] B. Worm and R. Paine, "Humans as a Hyperkeystone Species," *Trends in Ecology & Evolution* 31, no. 8 (2016): 600–607.

[5] Ibid.

[6] J. Syvitski, "Anthropocene: An Epoch of Our Making," *Global Change Magazine* (March 19, 2012): 12.

[7] ADWR, *Annual Report* (Phoenix, AZ: Arizona Department of Water Resources, 2016).

[8] City of New York, *Rules and Regulations for the Protection from Contamination, Degradation and Pollution of the New York City Water Supply and its Sources* (2010).

[9] P. Kareiva, M. Marvier, and M. McClure, "Recovery and management options for spring/summer chinook salmon in the Columbia River basin," *Science* 290, no. 5493 (2000): 977–979.

[10] R. Goldburg, and R. Naylor, "Future Seascapes, Farming, and Fish Farming," *Ecological Society of America* 3, no. 1 (2015): 21–28.

[11] S. Dhakal, "Urban energy use and carbon emissions from cities in China and policy implications," *Energy Policy* 37, no. 11 (2009): 4208–4219; S. Dhakal, "GHG emissions from urbanization and opportunities for urban carbon mitigation," *Current Opinion in Environmental Sustainability* 2, no. 4 (2010): 277–283; C. Kennedy, J. Steinberger, B, Gasson, Y. Hansen, T. Hillman, M. Havránek, D. Pataki, A. Phdungsilp, A. Ramaswami, and G. V. Mendez, "Greenhouse Gas Emissions from Global Cities," *Environmental Science & Technology* 43, no. 19 (2009): 7297–7302.; L. Parshall, K. Gurney, S. A. Hammer, D. Mendoza, Y. Zhou, and S. Geethakumar, "Modeling energy consumption and CO_2 emissions at the urban scale: Methodological challenges and insights from the United States," *Energy Policy* 38, no. 9 (2010): 4765–4782.

[12] T. Ming, R. de Richter, W. Lei, and S. Caillol, "Fighting global warming by climate engineering: Is the Earth radiation management and the solar radiation management any option for fighting climate change?" *Renewable and Sustainable Energy Reviews* 31 (2014): 792–834.

[13] A. Hommels, "Obduracy and urban sociotechnical change," *Urban Affairs Review* 35, no. 5, (2000): 649–676.

[14] A. Hommels, "Studying Obduracy in the City: Toward a Productive Fusion between Technology Studies and Urban Studies," *Science, Technology, & Human Values* 30, no. 3 (2005): 323–351.

[15] H. Corvellec, M. J. Z. Campos, and P. Zapata, "Infrastructures, lock-in, and sustainable urban development: The case of waste incineration in the Göteborg Metropolitan Area," *Journal of Cleaner Production* 50 (2013): 32–39.

[16] K. M. Faust, D. M. Abraham, and S. P. McElmurry, "Water and Wastewater Infrastructure Management in Shrinking Cities," *Public Works Management & Policy* 21, no. 2 (2016): 128–156.

[17] F. Krausmann, S. Gingrich, N. Eisenmenger, K.-H. Erb, H. Haberl, and M. Fischer-Kowalski, "Growth in global materials use, GDP and population during the 20th century," *Ecological Economics* 68, no. 10 (2009): 2696–2705.

[18] T. P. Hughes, "The Evolution of Large Technological Systems," in W. E. Bijker, T. P. Hughes, and T. J. Pinch, eds., *The Social Construction of Technological Systems: New Directions in the Sociology and History of Technology* (New York, NY: MIT Press, 1987), 51–81; B. Allenby, *The Theory and Practice of Sustainable Engineering* (Upper Saddle River, NJ: Pearson Prentice Hall, 2012); W. E. Bijker, T. P. Hughes, and T. Pinch, eds., *The Social Construction of Technological Systems: New Directions in the Sociology and History of Technology* (Cambridge, MA: MIT Press, 2012).

[19] R. Kurzweil, *Singularity is Near* (New York, NY: Viking, 2005); B. Allenby and D. Sarewitz, *The Techno-Human Condition* (Cambridge, MA: MIT Press 2011); G. E. Marchant, "The Growing Gap Between Emerging Technologies and the Law," in *The Growing Gap Between Emerging Technologies and Legal-Ethical Oversight: The Pacing Problem*, G. E. Marchant, ed. (Dordrecht, The Netherlands: Springer Netherlands, 2011), 19–33.

[20] A. Madrigal, "All the Promises Automakers Have Made About the Future of Cars," *The Atlantic* (July 7, 2017).

[21] D. Shoup, *The High Cost of Free Parking* (Chicago, IL: Planners Press, 2011).

[22] J. Flechas and J. Staletovich, "Miami Beach's battle to stem rising tides," *The Miami Herald*, (Oct. 23, 2015); J. Staletovich, "Beyond the high tides, South Florida water is changing," *The Miami Herald*, (Oct. 25, 2015); J. Flechas, "Miami Beach to begin new $100 million flood prevention project in face of sea level rise," *The Miami Herald* (March 23, 2017).

[23] P. C. D. Milly, J. Betancourt, M. Falkenmark, R. M. Hirsch, Z. W. Kundzewicz, D. P. Lettenmaier, and R. J. Stouffer, "Stationarity Is Dead: Whither Water Management?" *Science* 319, no. 5863 (2008): 573–574.

[24] J. Park, T. P. Seager, and P. S. C. Rao, "Lessons in risk- versus resilience-based design and management," *Integrated Environmental Assessment and Management* 7, no. 3 (2011): 396–399.

[25] M. Chester and B. Allenby, "Toward Adaptive Infrastructure: Flexibility and Agility in a Non-Stationarity Age," *Sustainable and Resilient Infrastructure* 3 (2018): 1–15.

[26] H. von Blottnitz and M. A. Curran, "A review of assessments conducted on bio-ethanol as a transportation fuel from a net energy, greenhouse gas, and environmental life cycle perspective," *Journal of Cleaner Production* 15, no. 7 (2007): 607–619; C. de Fraiture, M. Giordano, and Y. Liao, "Biofuels and implications for agricultural water use: blue impacts of green energy," *Water Policy* 10, no. S1 (2008): 67–81; T. Searchinger, R. Heimlich, R. A. Houghton, F. Dong, A. Elobeid, J. Fabiosa, S. Tokgoz, D. Hayes, and T.-H. Yu, "Use of U.S. croplands for biofuels increases greenhouse gases through emissions from land-use change," *Science* 319, no. 5867 (2008): 1238–1240; P. W. Gerbens-Leenes, A. Y. Hoekstra, and T. van der Meer, "The water footprint of energy from biomass: A quantitative assessment and consequences of an increasing share of bio-energy in energy supply," *Ecological Economics* 68, no. 4 (2009): 1052–1060.; T. W. Simpson, L. A Martinelli, A. N. Sharpley, and R. W. Howarth, "Impact of Ethanol Production on Nutrient Cycles and Water Quality: The United States and Brazil as Case Studies," in *Biofuels: Environmental Consequences and Interactions with Changing Land Use* (Ithaca, NY: Cornell University Library's Initiative in Publishing, 2009), 153–167; T. W. Hertel, A. A. Golub, A. D. Jones, M. O'Hare, R. J. Plemin, and D. M. Kammen, "Effects of U.S. Maize Ethanol on Global Land Use and

Greenhouse Gas Emissions: Estimating Market-mediated Responses," *BioScience* 60, no. 3 (2010): 223–231; M. Lagi, Y. Bar-Yam, K. Z. Bertrand, and Y. Bar-Yam, "The Food Crises: A Quantitative Model of Food Prices Including Speculators and Ethanol Conversion," *Social Science Research Network* (2011).

[27] International Energy Agency, *Global EV Outlook 2017: Two Million and Counting* (Paris, France: IEA Publications, 2017).

[28] H. Corvellec, M. J. Z. Campos, and P. Zapata, "Infrastructures, lock-in, and sustainable urban development: The case of waste incineration in the Göteborg Metropolitan Area," *Journal of Cleaner Production* 50 (2013): 32–39.

[29] Apple, *Environmental Responsibility Report* (Cupertino, CA: Apple, 2017).

[30] "Protein innovation: Cargill invests in cultured meats," Cargill (2017).

[31] H. Simon, *Administrative Behavior: A Study of Decision-making Processes in Administrative Organization* (Basingstoke, UK: Macmillan Company, 1947); H. Simon, *Models of Man: Social and Rational* (Berkeley, CA: Wiley, 1957).

[32] M. Chester and B. Allenby, "Toward Adaptive Infrastructure: Flexibility and Agility in a Non-Stationarity Age," *Sustainable and Resilient Infrastructure* 3 (2018): 1–15.

[33] T. S. Kuhn, *The Structure of Scientific Revolutions* (Chicago, IL: University of Chicago Press, 1962).

[34] M. Chester and B. Allenby, "Toward Adaptive Infrastructure: Flexibility and Agility in a Non-Stationarity Age," *Sustainable and Resilient Infrastructure* 3 (2018): 1–15.

[35] A. Darby and R. Wilson, "Design of the SMART project, Kuala Lumpur, Malaysia," conference paper presented at the International Conference and Exhibition on Tunnelling and Trenchless Technology (March 7–9, 2006); N. K. Hing, D. N. Welch, and T. S. Giap, "Stormwater Management and Road Tunnel (SMART), a bypass solution to mitigate flooding in Kuala Lumpur city center," conference paper presented at the International Conference and Exhibition on Tunnelling and Trenchless Technology (March 7–9, 2006).

[36] M. Chester and B. Allenby, "Toward Adaptive Infrastructure: Flexibility and Agility in a Non-Stationarity Age," *Sustainable and Resilient Infrastructure* 3 (2018): 1–15.

[37] E. Limer, "The Tech that Runs the NYC Subway is Positively Ancient," *Popular Mechanics,* (July 30, 2015); E. G. Fitzsimmons, "Key to Improving Subway Service in New York? Modern Signals," *The New York Times* (May 1, 2017).

[38] S. Arbesman, *Overcomplicated: Technology at the Limits of Comprehension* (New York, NY: Penguin, 2016).

[39] B. Allenby, *The Theory and Practice of Sustainable Engineering* (Upper Saddle River, NJ: Pearson Prentice Hall, 2012).

[40] A. Annila and S. Salthe, "Economies Evolve by Energy Dispersal," *Entropy* 12, no. 6 (2009): 606–633.

[41] J. Henrich and R. McElreath, "The evolution of cultural evolution," *Evolutionary Anthropology: Issues, News, and Reviews* 12, no. 3 (2003): 123–135.

7

THE CYBER FRONTIER AND INFRASTRUCTURE*

Mikhail Chester and Braden Allenby

The benefits of cybertechnologies integrated into infrastructure are becoming clearer. In 2017 the City of San Diego saw energy use drop by 60% after LEDs were installed downtown in conjunction with optical, auditory, and environmental sensors. In 2018 the Arizona Department of Transportation reported that more than a dozen wrong-way drivers were prevented from entering freeways by new thermal cameras and warning systems. California in late 2019 released an early warning system, providing residents with precious additional seconds to find safety before an earthquake.

The increasing integration of cybertechnologies into infrastructure is also creating vulnerabilities that we haven't ever experienced. A few days before Christmas in 2015 operators in the Prykarpattya Oblenergo electric utility of western Ukraine watched as their supervisory control and data acquisition (SCADA) system mouse pointer moved

* Reprinted with permission from M. V. Chester and B. R. Allenby, "Perspective: The Cyber Frontier and Infrastructure," *IEEE Access* 8 (2020): 28301–28310.

across the screen, no longer under their control, disabling substation after substation shutting down power across Ukraine. In 2017 hackers were able to access and transfer a casino's data using a vulnerability exploit in a wifi-connected fish tank sensor used to regulate water temperature, food, and water quality. In 2019 a ransomware attack brought the City of Baltimore's data management systems to a halt, suspending critical services related to real estate and communications.

Our increasingly connected systems are a new frontier for infrastructure, one that offers remarkable capabilities to deliver new or augmented services and lower costs, while on the other end it creates radically new vulnerabilities that have never been faced or even conceived. The integration of cyber and physical systems is accelerating. Yet the tools that we have at our disposal to manage this integration and the outlook that we have about what this integration means remain rooted in the past century.

The number of devices that are now connected is exploding, and infrastructure is part of the trend. Estimates vary but generally show acceleration of growth in both the number of connected devices and the amount of data being transferred. Devices and data are growing faster than the global population and number of internet users.[1] There are currently around 22 billion connected devices (approximately 3 per planetary citizen) with expectations of roughly 30 billion by 2022.

The growth in information traffic is outpacing that of devices. Mobile traffic has grown 17-fold between 2012 and 2017, and mobile devices are projected to average 10.7 gigabytes of data traffic per month by 2022, up from 2.3 gigabytes in 2017.[2] The amount and quality of data (e.g., video resolution) being transmitted is increasing.[3] Specific to infrastructure, the growth in machine-to-machine technologies (M2M) are of particular interest, with a projected 34% annual growth rate to 2022.[4] M2M refers to the direct communication between devices, which has been transitioning from closed network models to open. This allows devices to avoid communications hub and instead communicate directly with a centralized system or users (creating the potential for new technologies such as autonomous connected vehicle fleets). This category of interconnected devices has seen the largest growth, more than smart phones and personal computers. These devices are projected to drive much of the interconnectedness of smart cities and their infrastructure.

Viewing the embedding of smart technologies in infrastructure as simply an interconnectedness of systems is insufficient, if not irresponsible. The accelerating of the coupling may represent a singularity, a profound shift in the relationships between humans and their services.[5] It lays the groundwork for explosions of artificial intelligence, new capacities for services, radical changes in efficiency, and, with those, new vulnerabilities.

At the infancy of this shift, our comprehension of the implications of an accelerating cyberphysical world remains limited, and as such our ability to manage the implications and protect against vulnerabilities is likely woefully lacking. This unpreparedness has major implications for infrastructure managers and engineering education. It raises questions as to whether the next generation of leaders have the appropriate competencies to steer infrastructure as it transitions.

It's important to understand the context in which the acceleration of the interconnectedness between cyber and physical systems is happening. The demand for services delivered by infrastructure is one side of the story. Physical infrastructure systems (water, power, transportation, etc.) have largely been built to provide services that have for decades been relatively stable. We want water from a faucet the same way we did a century ago. How we demand electricity hasn't changed much from 1882 when Thomas Edison begin providing power through his Pearl Street Station to lower Manhattan. And over the past 70 or so years we (particularly those in the United States) have largely demanded automobility and its associated transportation infrastructure, which hasn't radically changed in technology (but certainly extent) in this time.

As such, the technologies that make up the backbone of our physical infrastructure systems have remained relatively stable for decades, if not centuries.[6] Certainly new technologies have been added, and efficiencies introduced, but water mains, pumps, transmission lines, transformers, and asphalt continue to dominate the core structure and functioning of these systems. If we were to bring Edison to today in a time machine he'd largely understand the power grid. But if we were to show Alexander Graham Bell a modern smart phone he'd be flummoxed by the black mirror. The acceleration of cyber technologies means that the cycle time (how quickly a past generation is replaced by a new generation) is now outpacing that of infrastructure. This is part of the challenge, working with cyberphysical systems that can't be treated as traditional coupled systems, given that cyber is cycling faster than the physical.

Concurrent with the technological change and increasing coupling of cyber and physical systems, there has been rapid acceleration in other fields, as well as social and political structures. Massive advances in computational power, data storage, and data analytics are driving advances in artificial intelligence and social media. At the same time we've seen a shift in military policy with a rise in asymmetric warfare strategies by nation states with weaker hardware, smaller armies, or less prepared armies that engage in cyberattacks to affect the strategic balance of power.[7] Nation-states have adopted explicit strategies of civilizational conflict which make targets of all of society's systems, from finance to infrastructure to health.[8]

The combination of rapid advancement of digital technologies, increasing interconnectedness of cyber and physical systems, different outlooks on humans, and differing approaches to warfare, represents a radically new paradigm, and infrastructure is at the center. We can't ignore this context as we design and manage infrastructure going forward.

Toward providing insights into the design and management of infrastructure in a future with potentially new demands for services, vulnerabilities, and relationships between people, the environment, and technologies, we explore the changing cyberphysical dynamics and its implications. We start by exploring technological acceleration theory and what that means for infrastructure. Next, we describe how transitions from physical to cyberphysical infrastructure will create new capabilities along with vulnerabilities. We consider the changing relationship between people and their services as mediated by infrastructure, as we accelerate the cyber integration of physical systems. We conclude by recommending how infrastructure education and management must shift from models that emphasize systems as they've traditionally existed to systems that will be controlled by cyber technologies.

We discuss three coupled but conceptually different systems: 1) cyber and information and communication technology (ICT) infrastructure; 2) physical infrastructure, which increasingly includes ICT functionality and technology; and 3) the "institutional context" of infrastructure (including education and management). We don't view this article as an exhaustive exploration or summary of all of the issues relevant to cyberphysical systems, but more as an effort to elucidate new thinking about the rapidly changing relationship between technologies, infrastructure, and people.

Accelerating Intelligence

In 1999 the futurist Ray Kurzweil noted that many technologies tend to grow exponentially and as such the twenty-first century can be expected to yield 20,000 historical years of relative progress.[9] He branded this phenomenon the Law of Accelerating Returns, which if true is accelerating humankind toward technological change so radical and profound that we cannot comprehend the implications.

The Law of Accelerating Returns is a theory of change acceleration, which in general describes the increasing rate of technological progress that ultimately results in profound social and cultural change; many theories of change acceleration exist.[10] While technology has always moved forward, the rate at which technology has changed up until recent times has mirrored population growth, meaning that essentially all of the world's population remained at subsistence levels of production and consumption.[11]

But technological change is now increasingly exponentially, representing a new paradigm for humans and the systems they operate.[12] This acceleration creates remarkable new opportunities, and also hazards and vulnerabilities, and implies a future that is difficult to meaningfully comprehend. How long such an acceleration can proceed has been the subject of much debate[13] However, as we look forward at the coming century there is accumulating evidence to warrant a critical examination of the implications of technological acceleration.[14] Technological acceleration is attributed to positive feedback loops, and trends of increasing integration of cyber into physical systems raise questions of whether the perceived benefits of cyber result in integration in infrastructure that thereby changes services and vulnerabilities creating new cyber-integration demands.

Maturation of Cyber Technologies

Infrastructures have, for decades, operated as either purely physical systems or with limited and often isolated computing capabilities, frequently included in system design as mechanical devices (inertial and centrifugal governors on steam engines might be regarded as a form of computational device, for example). Sensors, software, and digital controls (including SCADA) have been increasingly used since the latter half of the twentieth century, but these digital systems largely functioned to augment the core underlying infrastructure, which—for decades if not longer—have been largely driven by physical systems and hardware.

But recent advances in hardware and software are increasing the accessibility, usability, and affordability of cyber technologies. Computer scientist Ragurathan Rajkumar provides a useful synthesis of the factors that are pushing and pulling cyber technologies leading to their ubiquity.[15] Sensors are now available to measure the properties at nano to macro scales. Actuators have become ubiquitous (again across scales). Alternative energy sources are maturing. Satellite and wireless communications are available across the globe, and internet connectivity is growing. At the same time, computing and storage capabilities are improving, and appearing in ever smaller form factors. Demand for these technologies is also growing. Building and environmental controls, critical infrastructure monitoring, process control, factory automation, healthcare, aerospace, and defense are all advancing cyberphysical systems as industries strive for radical new capabilities and efficiencies.

The result is a new paradigm of infrastructure that includes hardware, software, firmware, and wetware (people) integrated into new techno-human infrastructures. These systems are now smart and connected, delivering the ability to measure system, natural, and human dynamics in ways that weren't feasible a short time ago. They are able to generate, see, and make sense of massive data streams (often using integrated AI and human capabilities), and send data to users in real time, in ways that heretofore have not been possible. This changing paradigm will shift how we interact with infrastructure and what we ask infrastructure to do.

The proliferation of lower cost, smaller, more efficient, and more powerful computing technologies coupled with data transfer, storage, and management technologies, and supported by emerging techniques (including AI) to make sense of voluminous and federated datasets represents profound new capabilities and efficiencies for physical systems.[16] This confluence of technologies represents an important transition period for infrastructure. Prior to this maturation we typically think of infrastructure as largely "dumb" physical systems.[17] The proliferation of the internet and augmentation of communication capacities (including bandwidth and communication protocols) has resulted in radical new possibilities for physical systems. These technologies represent new capabilities for how we understand and interact with natural and human systems.

Benefits of Cyberphysical Systems

The integration of cyber and physical systems creates new capabilities that didn't exist before. But it is what these capabilities enable that drive the accelerating integration of the systems. Prior to discussing infrastructure at broad scales it is useful to examine a parallel but smaller technology and its integration of cyber and physical systems: the automobile.

Until the 1950s automobile technologies had no cyber technologies; they were purely mechanical systems linked to each other via other mechanical systems or controlled through the cognitive capacities of the driver. The first sensors integrated into cars simply alerted drivers to problematic conditions such as low oil pressure via a dashboard light.[18] Critical system functioning was controlled by valves and other mechanical devices that responded directly to the driver's input.

In 1968 Volkswagen introduced the first microchip into a car to control fuel injection and minimize emissions, a device now known as an electronic fuel injector (EFI). Today's EFIs take in readings from dozens of subsystem sensors, perform millions of calculations per second, and adjust the spark timing and how long the fuel injector is open, ensuring the lowest emissions and highest fuel economy.[19] By 1999 cars had dozens of microprocessors.[20]

Today, sensors, processors, cameras, accelerometers, and other technologies result in 65Mb/s of data transferred throughout a vehicle, roughly 2 miles of cabling within the vehicle, and 280 connections to manage power and that data.[21] It's naive to think that pushing a gas pedal directly engages the engine. Instead a computer determines—based on your past behavior, environmental conditions, and readings from the vehicle—how to give you the best ride.

But this is just one scale of the system. Navigation software (e.g., Google Maps) now takes into account thousands of other drivers and routes you based not simply on the shortest travel time in an unloaded network but with consideration of how all other users of the system are traveling. Hybrid electric vehicles can learn your frequent destinations and automatically switch to electric power as you approach those destinations, thereby saving gas.[22] The integration of cyberphysical systems across scales as they relate to the automobile, the efficiencies they introduce, and the new capabilities radically alter our relationship with mobility services.

Cyberphysical infrastructure allow for new capabilities to optimize systems across broad scales and time frames, generate new efficiencies, and create multifunctionality where it didn't exist before. Fundamentally, the integration of cyber into physical systems creates new cognitive capacity about the system by shifting it toward relying more critically on information. New insights are created about not only the internal functioning and relationships between subsystems but also the demands (needs) being placed for the services. New optimization techniques are created with the integration of cyber creating the potential for efficiency gains. Sensors that detect ambient light can be used to control whether traffic lights are on or off, and their intensity when they're on, thereby reducing the need for electricity. Real-time information driving forecasts for electricity demand (power is perhaps the most historical major cyber-physical infrastructure) allows operators to deploy supply as needed. And in the case of mobility, a large scale connected vehicle fleet (possibly through Google Maps and emerging vehicle-to-vehicle communication technologies) offers the potential to shave peak demand, thereby reducing the need for new infrastructure and changing the kind of infrastructure that transportation engineers need to think about. Parking lots become less important; charging stations become more important.

These are possibilities that we can comprehend. With the deployment of artificial intelligence into cyberphysical systems the capabilities with full autonomy and humans not in control are beyond comprehension. But even today and in the near future, before the advent of fully autonomous AI, we must recognize that each new capability brings with it the potential for vulnerabilities and exploits.

Vulnerabilities

With great promise comes the potential for radically new vulnerabilities, the likes of which we have never seen with infrastructure. These vulnerabilities arise not simply because new exploits are created, but are largely due to the new capabilities for exploiting operators, users, control systems, distributed software, and hardware. These vulnerabilities arise at a time when cyberattack tools have become available to low-expertise hackers and nation-states have established and tested strategies and devoted resources for asymmetric warfare.

Taxonomies of Threats

To understand how cyber threats emerge in infrastructure a taxonomy is helpful. Many taxonomies exist for cyber threats, differing depending on the phase of the hacking process (data collection, storage, processing, etc.), target, actors, methods, techniques, or capabilities.[23] There is no preeminent taxonomy for threats to cyber-infrastructure systems.

It is helpful to first take a perspective within an infrastructure risk model. The National Institute of Standards and Technology's Guide for Conducting Risk Assessments provides a helpful model for framing risk factors for infrastructure management. Cyber threat taxonomies can map to the risk processes in NIST's model, providing a roadmap for analyzing threats as they relate to infrastructure. Starting with the threat source, taxonomies describe the types of actors involved in attacks, including professional criminals, state actors, terrorists, cyber vandals, hacktivists, internal actors, and cyber researchers.[24] They may focus on the threat event and techniques used, including degree of automation, exploited weakness, source address validity, possibility of characterization, attack rate dynamics, impact on the victim, victim type, and persistence of agent.[25] The attack vector, vulnerability, and exploit have also been the focus of taxonomies.

Analysts Simon Hansman and Ray Hunt catalog and map the types of attacks to targets (including hardware and software exploits) and the corresponding vulnerabilities.[26] Security researchers Charles Harry and Nancy Gallagher focus their taxonomy on the impacts of attacks by describing the outcomes of disruptions of operations and illicit acquiring of information.[27] Figure 1 shows how organizational risk is a result of different sources of attacks, event types, impacts, and vulnerabilities.

Although the level of sophistication and number of attacks has increased over time, the intruder expertise has decreased. This trend reflects the growing availability of tools to cyberattackers. While in the past considerable expertise and resources were needed to conduct an attack, it is becoming more and more common for a small number of expert hackers to make their tools available to a broader community of novice hackers. This growing body of cybercriminals has the capability to deploy an arsenal at ever-increasing scales, diversity, and sophistication with increasingly devastating effects.[28] This trend reflects a new reality of cyberattacks.

As National Intelligence Director James Clapper testified, "Rather than a 'Cyber Armageddon' scenario that debilitates the entire U.S. infrastructure, we envision something different. We foresee an ongoing series of low-to-moderate level cyberattacks from a variety of sources over time, which will impose cumulative costs on U.S. economic competitiveness and national security."[29] The ongoing low- to moderate-level attacks may reflect a Death by 1000 Cuts civilizational conflict strategy,[30] or simply that the vulnerabilities inherent in today's cyberdesigns attract an ever-increasing number of unrelated attacks.

Figure 1. Risk Model, Key Risk Factors and Associated Exemplary Taxonomies[31]

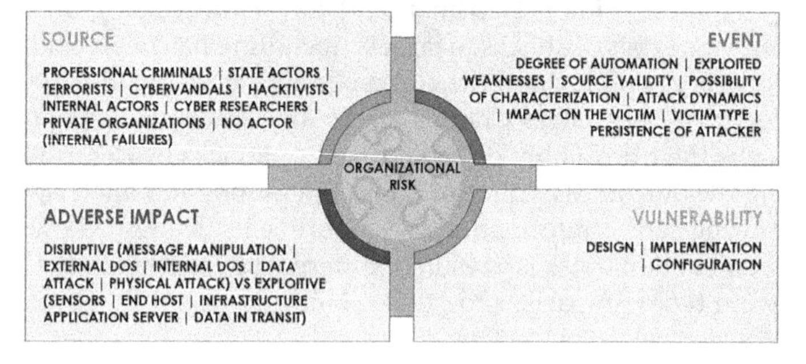

The level of sophistication of the attacker directly informs the strategies that should be developed when preparing for a cyberattack (Figure 2). Conventional threats include cyber vandalism and incursion, often involving disgruntled or suborned insiders, denial-of-service attacks, and hackers who have obtained legitimate user credentials.[32] Conventional threats can be approached by practice-driven risk management strategies that largely focus on basic hygiene and critical information protection (protocols for password changes, software updates, hardware updates, software installations, limiting users, and backing up data).

Figure 2. Sophistication of Cyber Threats, Risk Management Strategies, and Where in a Typical Infrastructure Divisional Bureaucracy Responsibility for Managing the Risk May Lie[33]

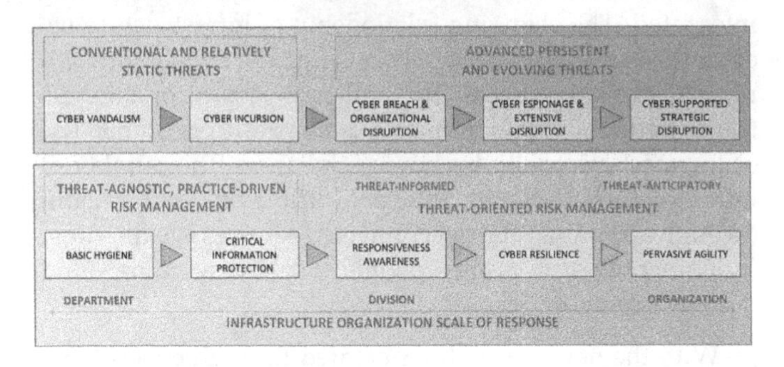

Advanced threats represent cyber adversaries that learn and evolve, such that compliance and good-practice driven strategies are insufficient and new competencies and threat-specific knowledge are needed (actors capable of such sophisticated cyber campaigns, such as Russia's Cozy Bear, are known as Advanced Persistent Threats).[34] While conventional threats can often be handled by a properly trained IT department, advanced threats may require sustained and directed resources for cybersecurity and management of corresponding initiatives, and appropriate staff, tools, and strategic planning. Agility may be required to consider the goals of attackers, the techniques the attacker may use, and the appropriate anticipatory and reactive responses an organization can deploy to protect itself.

Complexity and Vulnerability

As our infrastructure systems evolve toward greater complexity, in many ways defined by the increasing coupling of cyber and physical systems, vulnerabilities will need to be managed differently. We define infrastructure complexity here as the changing technical, environmental, and social context that engineers and managers must navigate to deliver and evolve services.[35] What is particularly interesting about the complexity associated with infrastructure is the speed and scale of which other systems are being integrated. The changing relationship of infrastructure users with the systems they rely on (e.g., the availability and price of parking spaces, the real-time arrival of the next transit vehicle, how to reroute to avoid traffic, the timed use of low-cost electricity by home appliances, the number of infrastructure elements that are offline in their region) through apps and internet connected services is exploding, fundamentally altering people's understanding and thereby use of services.[36]

With the new possibilities created through cyberphysical systems comes vulnerabilities and exploits that didn't exist before, some of which transcend the cyberphysical system. It's possible to conceive of cyberattackers no longer needing to target the cyberphysical system itself, but instead conditioning operators with targeted disinformation. Most attacks on infrastructure occur from within, generally disgruntled employees with internal access.[37] In 2006 engineers sabotaged intersection controls in Los Angeles, and in 2000 an ex-employee disabled critical SCADA systems with the hopes of being re-hired to fix the problem.[38]

With new means for engaging with these operators, for instance through social media, we can conceive of a new method for inciting sabotage without directly engaging with the infrastructure. In 2019 utility operators were targeted with emails impersonating their accreditation society baiting them to open malware attachments masked as notifications that their professional credentials were being revoked.[39] The attachments contained the LookBack virus that would give the cyberattackers access to the utility's systems.

Another vulnerability that is receiving considerable attention is the controlling of outgoing information about an attack to distort facts and condition a particular response. Reflexive control (the means of conveying to a partner or an opponent specially prepared information to incline him to voluntarily make the predetermined decision desired by the initiator of the action), a principle developed by Russia since the 1960s, is particularly well-suited for the hyperconnected and information rich era.[40] The case of the 2015 Ukrainian power grid cyberattack, for example, was part of a broader Russia strategy that involved denial of service attacks, disinformation campaigns (including social media, mass media, and internet trolls), and energy diplomacy (involving coercion that forced Ukraine to pay market prices for oil and gas), that together sowed disinformation across international outlets. The strategy allowed Russia to deploy minimal physical forces, thereby staying below the threshold for international intervention, while achieving their objective of stopping a revolution that threatened to overturn the pro-Russian administration.[41]

The possibilities for impact are no longer limited to the systems themselves, but span the interconnected systems in which our technical systems function. A challenge remains that those who understand the threat landscape and the complex tools being deployed are largely disconnected from those making day-to-day decisions about infrastructure. While in the United States the National Institute of

Standards and Technology and Department of Homeland Security issue valuable guidelines and recommendations for how to prepare for and protect against cyberattacks;[42] the reality of infrastructure at the ground level is one of limited resources and governing institutions that are structured to operate toward reliability principles that in many ways are designed to deliver services as they've been delivered in the past (and the existing engineering education structure reflects this).

Cyberwarfare Norm

That cybersecurity has become a major challenge for engineered systems is neither new nor particularly surprising. The roots of the challenge lie deep in recent geopolitical history. Partially because the United States was the strongest country left standing after World War II and the collapse of the Soviet Union, and partially because defense expenditures by the United States have consistently been far greater than those of any rival, the conventional military forces of the United States are generally understood to be stronger than those of any other power.[43]

This dominance has driven adversaries, especially state adversaries such as Russia and China, to adopt asymmetric warfare strategies that redefine conflict away from traditional military engagement to longer term "civilizational conflict," which among other things elevates information warfare, disinformation and subversion techniques, and weaponized narrative to priority attack mechanisms.[44] In perhaps the most cited military strategy article of the past decade, General Valery Gerasimov, Chief of the General Staff of the Russian Federation, notes that in the twenty-first century there has been "a tendency toward blurring the lines between the states of war and peace," and that "a perfectly thriving state can, in a matter of months and even days, be transformed into an arena of fierce armed conflict, become a victim of foreign intervention, and sink into a web of chaos, humanitarian catastrophe, and civil war."

Writing before the successful Russian invasion of Crimea and eastern Ukraine in 2014, General Gerasimov emphasizes:

The very "rules of war" have changed. The role of nonmilitary means of achieving political and strategic goals has grown, and, in many cases, they have exceeded the power of force of weapons in their effectiveness.... The focus of applied methods of conflict has altered in the direction of the broad use of political, economic, informational, humanitarian, and other nonmilitary measures – applied in coordination with the protest potential of the population. All this is supplemented by military means of a concealed character, including carrying out actions of information conflict and the actions of special operations forces. The open use of forces – often under the guise of peacekeeping and crisis regulation – is resorted to only at a certain stage, primarily for the achievement of final success in the conflict.[45]

Russia is not alone in developing civilizational conflict strategies as an asymmetric response to American conventional dominance. Shocked by the success of allied forces in Desert Storm (1990–1991), Chinese strategists have developed a strategy of "Unrestricted Warfare" that contemplates conflict across the entire domain of a civilization, from financial markets to all forms of infrastructure:

There is reason for us to maintain that the financial attack by George Soros on East Asia, the terrorist attack on the U.S. embassy by Usama bin Laden, the gas attack on the Tokyo subway by the disciples of the Aum Shinri Kyo, and the havoc wreaked by the likes of Morris Jr on the Internet, in which the degree of destruction is by no means second to that of a war, represent semi-warfare, quasi-warfare, and sub-warfare, that is, the embryonic form of another kind of warfare.[46]

Iran, North Korea, and others are following in Russian and Chinese footsteps, although not as part of such a structured and formal geopolitical conflict strategy.

The complacency of academic engineering education institutions in light of active cyberwarfare directed at essentially all engineered systems within American, and Western, society is remarkable—and untenable. Engineering students in disciplines including civil and environmental, biomedical, and industrial engineering are taught to include ever-more advanced sensor, computing, communication, and data processing systems in their designs because of concomitant dramatic improvements in function and efficacy. But they are taught next to nothing about information and cybersecurity, both because of the inertia of engineering curricula to any proposed change, and because their professors were never trained in the subject, are not versed in it, and completely fail to perceive, much less understand, relevant geopolitical shifts.

The result is that American engineering education is optimally designed to create a generation of engineering professionals who will, among other things, unknowingly design ever more vulnerability and frailty into the built environment and infrastructure systems that are critical to our society. We should recognize that the integration of cyber technologies into infrastructure is altering the relationships between people and their services.

Humans, Their Services, and the Environment, Mediated by Software

Edwin Hutchin's 1995 book *Cognition in the Wild* describes, through the lens of U.S. Navy pilots and sailors, the differences in cognitive approaches between individuals with no technology (the first sailors) and groups with technology.[47] Hutchins argues that cognition in modern society is composed of multiple agents and their technologies. While sailors on a modern Navy vessel cannot necessarily navigate like early sailors with no technology, they are able to accomplish remarkably more, by compartmentalizing tasks, communicating effectively, and utilizing technology.

Technology creates new opportunities for understanding the world around us, and as it accelerates is likely to create radical new relationships between people and their environments.

The rapid integration of cyber technologies into infrastructure and the implications for how humans interact with and demand services may represent a fundamentally new relationship that remains difficult if not impossible to comprehend. Whereas in the past new technologies often represented new capabilities and efficiencies, the hyperconnected and information-driven reality represents a radical change in how we see and experience the world. And artificial intelligence that mediates our interactions with other people, information, and services is positioned to fundamentally alter human experience. Infrastructure is at the center of this change.

Physical infrastructure systems will remain the backbone for cybertechnologies, but how they're used is poised to radically change. Several key dynamics may reshape our relationships with infrastructure:

- *Physical Systems as the Cyber Backbone*: Despite shifts from hardware to software functionality that reduces the need for physical assets,[48] core physical systems will be needed to enable information transfer, analytics, and storage. And who controls the core physical systems will be strategically positioned, both economically and politically (see Google and Facebook's efforts to deploy fiberoptic lines around the world and recent concern over 5G hardware security).[49]
- *Insights into Infrastructure Services*: Next, people are gaining and will continue to gain new insights about infrastructure that they didn't have before, thereby changing how they demand infrastructure services. The advent of smartphones created an industry of location tracking and traffic analysis firms that now deliver products and insights to travelers about the conditions

of roads, how to route to minimize delays, and how to change their travel behaviors to reduce trip times.[50] While still in its infancy, the possibilities of software making sense of the complexity of the transportation system has remarkable implications for how we use the system based on how the software understands it. Imagine similar insights and software-driven intelligence behind water and energy use, for example. And we are already heavily debating and seeing the implications of such intelligence driving how we consume news and media.[51]

- *Evolving Demands for Infrastructure Services*: While it's easy to imagine how new and improved information can make our interactions with infrastructure more efficient (e.g., saving us travel time or managing our appliances to run at low-cost electricity times of day), it's likely that the possibilities offered by cyber technologies will result in demands for new services. The emergence of car, bicycle, and scooter sharing, which is resulting in major changes to how people travel in many major cities,[52] would not have been possible without smartphones and cellular networks. Furthermore, combining modalities with autonomous vehicles means that you end up needing to redefine the urban transportation network completely.
- *Adaptive Capacity*: The integration of sensing technologies coupled with analytical capabilities and software-based intelligence is likely to create new adaptive capacities for infrastructure. Sensors of various forms that can detect the conditions of assets in the system, both in terms of structure and function, are already being deployed and utilized in new ways. This information will likely drive algorithms that make sense of the overall state of the system, and decisions about how to manage assets to ensure integrity and efficiency. Imagine a SCADA system deciding to triage a portion of a water distribution network where a pipe is expected to fail to

ensure that a cascading failure does not ensue. This capability will likely increase the agility and flexibility of infrastructure services to meet rapid changes in conditions and respond to hazards. Google Maps may already be showing us this adaptivity, by rerouting users with considerations of larger systems dynamics when there is a traffic accident.

These changes represent just a few of the possibilities of how cyber technologies may change our relationship with infrastructure. Preparing infrastructure managers and engineers for these shifts is critical to ensuring the integrity and safety of cyberphysical systems. As the technologies that define infrastructure change, so must education and governance for these systems.

Preparing for Cyberinfrastructure

Several critical and immediate efforts are needed to ensure that the integration of cyber-infrastructure results in systems that continue to support society's needs and are safe and secure. While there are certainly hardware and software changes that are needed, we focus these efforts on the institutional management of infrastructure and the training of future managers.

The training around integrated cyberphysical systems at universities is essentially non-existent and should immediately be developed as a core competency, a Fifth Column that can change the status quo of how we view and manage infrastructure.[53] Engineers, architects, planners, and other infrastructure managers will still need knowledge around fundamentals of design principles, underlying science, and operations. However, they will need to be trained with new competencies that support a new norm for infrastructure, one where systems are increasingly focused on information management.[54]

Currently, disciplines such as civil, environmental, and mechanical engineering; planning; and architecture (domains largely responsible for the physical systems) mostly train independently of computer science, computer engineering, information sciences, and military or security domains. This fragmentation of knowledge is likely to lead to unintended consequences, both in the relevancy of disciplines and who and what decides how infrastructure services are managed. Cyber technology, information management, and security must become central to the training of infrastructure managers.

Cyber security competencies must become central to the training of infrastructure managers. To attempt to manage infrastructure today without such training is an ethical and professional failure, particularly in light of the increasing cyberattacks on infrastructure. New managers must have at least basic competencies to know why different actors might want to target their systems, what techniques they can use to exploit vulnerabilities, and strategies that can be deployed to protect systems (Figure 2).[55]

It has become remarkable easy (both in terms of technology and cost) to layer new and connected technologies into old and new systems, without a comprehensive understanding of the implications, risk, and vulnerabilities. Infrastructure managers must be trained with the tools to understand how to vet hardware and software on devices, encrypt and secure communications, manage access to information, and thwart inside and outside attacks.

Infrastructure managers will need to develop roadmaps that guide the planning and development of their cyber systems into physical systems. This will require translating federal insights to their locales. It is difficult to find cyber planning and cybersecurity plans for state, regional, and local infrastructure agencies. These plans should immediately be developed and serve as a roadmap for how infrastructure agencies plan on integrating cyber into their systems and protect their systems against threats. Much of the cybersecurity literature identified was developed by federal agencies (namely NIST and DHS) and there is good reason to assume that intelligence agencies are also central to making sense of the challenge.

However, when it comes to day-to-day decisions about infrastructure assets, limited guidance exists. This information is sorely needed. It should be specific to region (considering local needs and hazards), describe threats across scales (from foreign to local actors), guide managers in how to access vulnerabilities in hardware and software, and provide strategies for protecting systems.

A remarkably difficult challenge will be steering infrastructure as artificial intelligence comes online. Software developers that are developing artificial intelligence appear to be doing so largely independent from those that design and manage infrastructure. The implications of this lack of coordination are very unclear. Will the software manage services in ways that infrastructure managers hadn't intended? Will it drive infrastructure development in an unplanned direction? Will it monopolize resources beyond the capacity of the system?

For every question that we think of there's probably two more that are beyond our comprehension, given the complexity of an AI-managed system and its potential for restructuring how we understand and interact with human systems. What is clear is that the tools and techniques that we currently train and deploy are badly out of date, and that the acceleration of the integration of cyber and physical systems is not likely one that we will be able to control. Instead, we'll need to accept that our new role is one of understanding and guiding the emerging complexity.

Conclusion

New approaches to how we think about goals and structure of infrastructure, what those systems do, and how they are operated are immediately needed to ensure that societal needs are met into the future. The cyber technologies that are increasingly integrated with physical systems are being developed faster than the infrastructure, resulting in an increasing mismatch between the new capabilities delivered by the cyber technology and the obdurate backbone physical system's capabilities. This is likely to lead to unintended consequences in how infrastructure are used and their reliability.

Furthermore, the acceleration of technologies, their pervasive use, and a dearth of knowledge and training among infrastructure managers is creating major vulnerabilities that are already being exploited. Education of infrastructure managers must include cyber technology. The growing complexity of human systems and their relationships with natural and social systems appears to be accelerating. The sooner we accept that the approaches we use to manage the core infrastructure systems that support human activities are rooted in the past century, the sooner we can reinvent infrastructure management for the coming centuries.

Notes

[1] "Cisco Visual Network Index: Forecast and Trends, 2017-2022," Cisco Systems (2019).

[2] "Cisco Visual Networking Index: Global Mobile Data Traffic Forecast Update, 2017-2022," Cisco Systems (2019).

[3] "Cisco Visual Network Index: Forecast and Trends, 2017-2022," Cisco Systems (2019); "Ericsson Mobility Report," Ericsson (2017).

[4] "Cisco Visual Network Index: Forecast and Trends, 2017-2022," Cisco Systems (2019).

[5] R. Kurzweil, "The Law of Accelerating Returns," in *Alan Turing: Life and Legacy of a Great Thinker*, C. Teuscher, ed. (Berlin, Germany: Springer Berlin Heidelberg, 2004), 381–416.

[6] M. V. Chester and B. Allenby, "Toward adaptive infrastructure: flexibility and agility in a non-stationarity age," *Sustainable and Resilient Infrastructure* 4, no. 4 (2018): 173–191.

[7] M. V. Chester and B. Allenby, "Toward adaptive infrastructure: flexibility and agility in a non-stationarity age," *Sustainable and Resilient Infrastructure* 4, no. 4 (2018): 173–191; J. Fritz, "How China will use cyber warfare to leapfrog in military competitiveness," *Culture Mandala: The Bulletin of the Centre for East-West Cultural and Economic Studies* 8, no. 1 (2008): article 2.

[8] B. R. Allenby, "The paradox of dominance: The age of civilizational conflict," *Bulletin of the Atomic Scientists* 71, no. 2, (2015): 60–74.

[9] R. Kurzweil, "The Law of Accelerating Returns," in *Alan Turing: Life and Legacy of a Great Thinker*, C. Teuscher, ed. (Berlin, Germany: Springer Berlin Heidelberg, 2004), 381–416.

[10] R. Kurzweil, "The Law of Accelerating Returns," in *Alan Turing: Life and Legacy of a Great Thinker*, C. Teuscher, ed. (Berlin, Germany: Springer Berlin Heidelberg, 2004), 381–416; P. T. de Chardin, *The Phenomenon of Man* (New York, NY: Harper Collins, 2008); R. L. Coren, *The Evolutionary Trajectory: The Growth of Information in the History and Future of Earth* (Netherlands: Taylor & Francis, 2003); L. Nottale, J. Chaline, and P. Grou, *Les arbres de l'évolution: univers, vie, sociétés* (France: Hachette littératures, 2000).

[11] G. Clark, *A Farewell to Alms: A Brief Economic History of the World* (Princeton, NJ: Princeton University Press, 2007); N. Rosenberg and L. E. Birdzell, *How the West Grew Rich: The Economic Transformation of the Industrial World* (USA: Basic Books, 1986).

[12] J. Syvitski, "Anthropocene: An Epoch of Our Making," *Global Change*, 78 (2012): 12–15; B. Nagy, J. D. Farmer, J. E. Trancik, and J. P. Gonzales, "Superexponential long-term trends in information technology," *Technological Forecasting and Social Change* 78, no. 8 (2011): 1356–1364.

[13] J. Rennie, "Ray Kurzweil's Slippery Futurism," *IEEE Spectrum: Technology, Engineering, and Science News* (Nov. 29 2010).

[14] B. Nagy, J. D. Farmer, J. E. Trancik, and J. P. Gonzales, "Superexponential long-term trends in information technology," *Technological Forecasting and Social Change* 78, no. 8 (2011): 1356–1364.

[15] R. Rajkumar, "A Cyber–Physical Future," *Proceedings of the IEEE* 100, no. Special Centennial Issue, (2012): 1309–1312.

[16] Ibid.

[17] B. Allenby, "Infrastructure in the Anthropocene: Example of Information and Communication Technology," *Journal of Infrastructure Systems* 10, no. 3, (2004): 79–86.

[18] R. Jurgen, *History of Automotive Electronics* (Warrendale, PA: Society of Automotive Engineers, 1998).

[19] "Computer Chips Inside the Car," Chips Etc. (2014).

[20] J. Turley, "Embedded Processors by the Numbers," *EETimes* (1999).

[21] A. Katwala, "Connected cars are 'driving microchip development,'" *Institution of Mechanical Engineers* (Aug. 31, 2017).

[22] M. Amick, "First drive: Lincoln's 2013 MKZ hybrid makes going green easy, but lacks luxury," *Digital Trends* (May 19, 2013).

[23] C. Harry and N. Gallagher, "Classifying Cyber Events: A Proposed Taxonomy," *Journal of Information Warfare* 17, no. 3 (2018): 17.

[24] M. de Bruijne, M. van Eeten, C. Hernández Gañán, and W. Pieters, "Towards a new cyber threat actor typology: A hybrid method for the NCSC cyber security assessment," Delft University of Technology, Netherlands (2017).

[25] J. Mirkovic and P. Reiher, "A Taxonomy of DDoS Attack and DDoS Defense Mechanisms," *Computer Communication Review* 34, no. 2 (2004): 39–53.

[26] S. Hansman and R. Hunt, "A taxonomy of network and computer attacks," *Computer Security* 24, no. 1 (2005): 31–43.

[27] C. Harry and N. Gallagher, "Classifying Cyber Events: A Proposed Taxonomy," *Journal of Information Warfare* 17, no. 3 (2018): 17.

[28] S. Hansman and R. Hunt, "A taxonomy of network and computer attacks," *Computer Security* 24, no. 1 (2005): 31–43; H. Lipson, "Tracking and Tracing Cyber-Attacks: Technical Challenges and Global Policy Issues," CMU/SEI-2002-SR-009, Carnegie Mellon University Software Engineering Institute (2002); National Research Council, *The Resilience of the Electric Power Delivery System in Response to Terrorism and Natural Disasters: Summary of a Workshop* (Washington, DC: National Academies Press, 2013).

[29] J. Clapper, "Statement for the Record: Worldwide Cyber Threats Before the House Permanent Select Committee on Intelligence" (Sept. 10, 2015).

[30] B. R. Allenby, "The paradox of dominance: The age of civilizational conflict," *Bulletin of the Atomic Scientists* 71, no. 2, (2015): 60–74.

[31] Based on C. Harry and N. Gallagher, "Classifying Cyber Events: A Proposed Taxonomy," *Journal of Information Warfare* 17, no. 3 (2018): 17; M. de Bruijne, M. van Eeten, C. Hernández Gañán, and W. Pieters, "Towards a new cyber threat actor typology: A hybrid method for the NCSC cyber security assessment," Delft University of Technology, Netherlands (2017); J. Mirkovic and P. Reiher, "A Taxonomy of DDoS Attack and DDoS Defense Mechanisms," *Computer Communication Review* 34, no. 2 (2004): 39–53; S. Hansman and R. Hunt, "A taxonomy of network and computer attacks," *Computer Security* 24, no. 1 (2005): 31–43; S. Hansman and R. Hunt, "A taxonomy of network and computer attacks," *Computer Security* 24, no. 1 (2005): 31–43; National Institutes of Standards and Technology, "Guide for Conducting Risk Assessments," 800-30r1 (2012); D. Bodeau and R. Graubart, "Cyber Prep 2.0: Motivating Organizational Cyber Strategies in Terms of Threat Preparedness," The MITRE Corporation (2017); and J. Howard and T. Longstaff, "A Common Language for Computer Security Incidents," Sandia National Laboratories, SAND98-8667 (1998).

[32] D. Bodeau and R. Graubart, "Cyber Prep 2.0: Motivating Organizational Cyber Strategies in Terms of Threat Preparedness," MITRE Corporation (2017).

[33] Adapted from D. Bodeau and R. Graubart, "Cyber Prep 2.0: Motivating Organizational Cyber Strategies in Terms of Threat Preparedness," The MITRE Corporation (2017).

[34] Ibid.

[35] M. V. Chester and B. Allenby, "Infrastructure as a wicked complex process," *Elementa: Science of the Anthropocene* 7, no. 1 (2019): 21.

[36] M. Sarwar and T. Rahim Soomro, "Impact of Smartphones on Society," *European Journal of Scientific Research* 98, no. 2 (2013): 216–226.

[37] A. Cardenas, S. Amin, B. Sinopoli, A. Giani, A. Perrig, and S. Sastry, "Challenges for Securing Cyber Physical Systems," Workshop on Future Directions in Cyber-Physical Systems Security, Department of Homeland Security, (July 23, 2009).

[38] J. Slay and M. Miller, "Lessons Learned from the Maroochy Water Breach," in *Critical Infrastructure Protection*, E. Goetz and S. Shenoi, eds. (Boston, MA: Springer, 2008), 73–82; T. Rid, "Cyber-Sabotage Is Easy," *Foreign Policy* (July 23, 2013).

[39] M. Raggi and D. Schwarz, "LookBack Malware Targets the United States Utilities Sector with Phishing Attacks Impersonating Engineering Licensing Boards," *Proofpoint Blog* (Aug. 1, 2019).

[40] T. Thomas, "Russia's Reflexive Control Theory and the Military," *Journal of Slavic Military Studies* 17, no. 2 (2004): 237–256.

[41] F. King, "Reflexive Control and Disinformation in Putin's Wars," Colorado State University (2018); R. Sprang, "Russia in Ukraine 2013-2016: The Application of the New Type of Warfare Maximizing the Exploitation of Cyber, IO, and Media," *Small Wars Journal* (2018).

[42] Joint Task Force Transformation Initiative, "Guide for Conducting Risk Assessments," Special Publication 800-30, National Institute of Standards and Technology (2012); "Department of Homeland Security's Cybersecurity and Infrastructure Security Agency," Department of Homeland Security (Nov. 20, 2018).

[43] B. Allenby, "In an Age of Civilizational Conflict," *Jurimetrics* 56, no. 4 (2016): 387–406; H. Kissinger, *World Order* (New York, NY: Penguin Press, 2014); S. McFate, *The New Rules of War: Victory in the Age of Durable Disorder* (New York, NY: HarperCollins, 2019); Q. Liang and W. Xiangsui, "Unrestricted Warfare," PLA Literature and Arts Publishing House, Beijing, China (1999).

[44] B. R. Allenby, "The paradox of dominance: The age of civilizational conflict," *Bulletin of the Atomic Scientists* 71, no. 2, (2015): 60–74; B. Allenby and J. Garreau, "Weaponized Narrative: The New Battlespace," New America Foundation and Arizona State University's Center on the Future of War (2017); P. W. Singer and E. T. Brooking, *LikeWar: The Weaponization of Social Media* (New York, NY: Houghton Mifflin Harcourt, 2018).

[45] V. Gerasimov, "The Value of Science is in the Foresight," *Military Review* (Jan.-Feb., 2016): 23–29.

[46] Q. Liang and W. Xiangsui, "Unrestricted Warfare," PLA Literature and Arts Publishing House, Beijing, China (1999).

[47] E. Hutchins, *Cognition in the Wild* (Cambridge, MA: MIT Press, 1995).

[48] W. Shih, "Does Hardware Even Matter Anymore?" *Harvard Business Review* (June 9, 2015).

[49] M. Burgess, "Google and Facebook are gobbling up the internet's subsea cables," *Wired Magazine* (Nov. 18, 2018); C. Bryan-Low, C. Packham, D. Lague, S. Stecklow, and J. Stubbs, "Hobbling Huawei: Inside the U.S. ware on China's tech giant," *Reuters* (May 21, 2019).

[50] D. Wang and D. R. Fesenmaier, "Transforming the Travel Experience: The Use of Smartphones for Travel," in *Information and Communication Technologies in Tourism 2013*, L. Cantoni and Z. Xiang, eds. (Berlin, Germany: Springer, 2013), 58–69; X. Hu, Y.-C. Chiu, and J. Shelton, "Development of a behaviorally induced system optimal travel demand management system," *Journal of Intelligent Transportation Systems* 21, no. 1 (2017): 12–25; S. Kim and B. Coifman, "Comparing INRIX speed data against concurrent loop detector stations over several months," *Transportation Research Part C: Emerging Technologies* 49 (2014): 59–72.

[51] Z. Tufekci, "Algorithmic Harms beyond Facebook and Google: Emergent Challenges of Computational Agency," *Colorado Technology Law Journal* 2 (2015): 203–218.

[52] R. Clemlow and G. Shankar Mishra, "Disruptive Transportation: The Adoption, Utilization, and Impacts of Ride-Hailing in the United States," UCD-ITS-RR-17-07, University of California Davis Institute of Transportation Studies (2017).

[53] P. W. Senge, *The Fifth Discipline: The Art & Practice of a Learning Organization* (New York, NY: Doubleday, 1990).

[54] B. Allenby, "5G, AI, and big data: We're building a new cognitive infrastructure and don't even know it yet," *Bulletin of the Atomic Scientists* (Dec. 19, 2019).

[55] D. Bodeau and R. Graubart, "Cyber Prep 2.0: Motivating Organizational Cyber Strategies in Terms of Threat Preparedness," MITRE Corporation (2017).

8

INFRASTRUCTURE AS A WICKED COMPLEX PROCESS*

Mikhail Chester and Braden Allenby

Herbert Simon, a pioneer of decision-making theory, posited that given the vast information that is needed to completely understand how to maximize one's benefit from a particular course of action, people instead "satisfice." That is, they will make a decision "which is good enough, rather than the absolute best."[1] Satisficing, a combination of satisfy and suffice, describes the process by which, in situations of wicked complexity where optimization techniques fail, we often settle on a course of action that is good enough and "deals with a drastically simplified model of the confusion that constitutes the real world."[2]

We use the term "optimization" to describe traditional engineering decision making approaches that emphasize quantification of performance with a focus on maximizing performance or efficiency while minimizing costs and meeting some desired level of service. Simon initially de-

---

* Reprinted with permission from M. V. Chester and B. Allenby, "Infrastructure as a wicked complex process," *Elementa* 7, no. 1 (2019): 21.

veloped the concept of satisficing at a time when commercial, industrial, and public sector organizations were restructuring post World War II to peacetime services. It had profound implications on how we view decision making, not as a process that uses complete and consistent information and preferences but instead as a "rational behavior that is compatible with the access to information and the computational capacities that are actually possessed by organisms."[3] With what appears to be growing complexity in the constraints imposed on how our infrastructure systems are designed and operated, it's necessary to examine how the role of engineering is changing from optimizing to satisficing.

What our infrastructure will need to do in the coming centuries is likely very different than today. We acknowledge that the term infrastructure can have broad and evolving meanings,[4] but in general refers to the human-made physical and institutional elements of systems that provide services.[5] In the Anthropocene, this could include many human systems, so to clarify the implications for the role of engineering, this paper will focus on civil infrastructure systems that, for example, provide water and energy, facilitate transportation, and provide information and communication capabilities.

Several long-term trends appear to be creating complexity for our infrastructure systems. First, our ability to change infrastructure at a meaningful pace appears to be constrained. The infrastructure that we rely on today shares many of the core design features from when these systems were initially conceived decades ago.[6] Because demand for the core services that infrastructures deliver hasn't changed for many civil systems, the physical forms that infrastructure takes and the institutions that support those forms have been arranged and connected with each other over such a long history that shifting to new forms of infrastructure appears a monumental challenge—we describe this effect as *lock-in*.

Second, as the scale and scope of human activities have grown, infrastructure systems have incorporated more and more of "nature." The dichotomy between the human-built and natural worlds is shrinking, thereby greatly increasing complexity, both technically and culturally.[7]

Third, increasingly rapid changes in technology are driving our infrastructure to deliver services that they weren't designed to deliver – a trend that will only increase in the future. This means that we need faster cycle times. Yet as per the first trend, our infrastructures appear to be increasingly locked-in, slowing down our ability to change them quickly. As the need for more rapid and complex infrastructure upgrades, redesigns, and construction accelerates, lock-in becomes a significant constraint and likely long-term trend.

Fourth, because civil infrastructures have persisted for so long there has been significant accretion – i.e., the accumulation, layering, and interconnectedness of technologies, institutions, rules, and policies – so that understanding the emergent behavior of infrastructure (particularly when perturbed) is now challenging if not impossible. Significant change now requires action across many different subsystems and their governing agencies. For example, modernizing the U.S. power grid requires the coordinated deployment, financing, and permitting of hardware and software across hundreds if not thousands of companies and agencies.

The combination of these four factors raises serious questions about whether rapidly changing demands, technologies, and perturbations (such as climate change or terrorist attacks) will affect our infrastructure's capacity to provide services. It is these concurrent trends that also demand a shift from optimization to satisficing as the critical engineering competency.

It is now widely recognized that we are at the dawn of the Anthropocene, the era of the human, where rapid environmental, technological, social, political, and even cultural changes, increasingly at regional and global scales, are likely to put new demands on our infrastructure. Yet our infrastructure remains mostly rigid, unable to change quickly.[8] Making infrastructure agile and flexible for the Anthropocene will require us to acknowledge and work with the fact that infrastructures are now wicked complex systems, and that satisficing has become the new normal for designing and managing these systems.

This has profound implications for infrastructure managers who have been trained to assess bounded design problems that can be optimized, rather than wickedly complex problems that must be satisficed. In this chapter we first explore how infrastructures have become wicked complex systems. We then discuss how this wicked complexity has led to a shift in how we design and operate infrastructure and the challenges this approach introduces when large and fast changes are needed. Lastly, we propose that engineering (including education) embraces the wicked complexity of infrastructure and the increasing satisficing that is needed to affect change.

Infrastructure Design through the Ages

Modern infrastructure systems are several millennia in the making. The Neolithic revolution saw many human cultures shift from hunter gatherer societies to farming and agriculture. During the Neolithic period (approximately 10000 to 2000 BC), deforestation, terraforming, crop domestication, and irrigation resulted in densely populated settlements. The era also witnessed increased specialization in activites, the establishment of social and sexual roles and hierarchies, and new outlooks on ownership.

During this time, in addition to farming and irrigation, dirt road and durable good technology developed.[9] Town systems by were common by the end of the Neolithic paving the way for the Classical era (800 BC to 600 AD) in Europe, and significant scientific, technological, and cultural advances in the Chinese and Islamic civilizations. During this time the Greco-Roman world developed paved roadways, aqueducts and water mills, and metal use for building construction, communications, and energy use. Driven by extensive trading networks, transportation technologies advanced in Chinese and Islamic empires. And in Asia there was considerable development of agricultural technologies and practices.

Through the Middle Ages (500 to 1500 AD), infrastructure and associated technology development accelerated. The horse harness, watermills, heavy plow, metallurgy, gunpowder, gothic architecture, and the caravel developed in Europe, much of which was borrowed and modified from elsewhere.[10] By the Industrial Era, infrastructure and technologies were rapidly changing.

The rate of change of infrastructure and technologies over the ages provides an important perspective on how much faster human systems are changing today than in the past. In the past century alone, the level of social, cultural, financial, and political complexity has exploded to the point where the rules by which we deploy infrastructure have drastically changed.[11]

Today's cities are integrative technologies in that they reflect the engineering, culture, and infrastructure of their various periods. Medieval cities, for example, had narrow, crooked streets crowded with non-standardized buildings that reflected the dominant transportation mode (walking and the occasional horse), and a lack of zoning or building regulations. The streets doubled as sewage infrastructure; the lack of toilets meant that chamber pots were used, and were simply emptied out of windows.[12] Slow modernization of urban systems, from sanitation to public transit, marked the eighteenth and nineteenth centuries, culminating in "high modernism."

High modernism as practiced by planners such as Robert Moses in New York, and Le Corbusier in urban design, was marked by a brutal technocratic elitism which knowingly ignored local context and culture in the interest of implementing "universalist" and "scientific" principles in urban design. As Robert Moses declared, "When you operate in an overbuilt metropolis, you have to hack your way with a meat ax.... I'm just going to keep right on building. You do the best you can to stop it" (as quoted by the philosopher Marshall Berman).[13] Critics such as Jane Jacobs and Charles Jencks derided high modernism as anti-human and hubristic, and favored "post-modern" city planning and architecture.[14]

And indeed top down high modernism was replaced by postmodernist bottom-up community activism, seen in such controversies as the defeat of the proposed Greenpoint-Williamsburg waste incinerator in Brooklyn and the London Docklands proposal, both in the 1990s.[15] The resulting power of activists and communities, while it accords with modern sentiment in many ways, makes many large projects such as Boston's "Big Dig" major political efforts, and can allow small but vocal minorities to stifle projects and infrastructure that the larger public good or urban community requires. In other words, the postmodern political environment of today makes wicked complexity integral to virtually any significant infrastructure project, a point that today's engineers—and the professors that teach them—cannot ignore.

Modern infrastructure is thus not a technical problem, it's a wickedly complex problem. Going forward, we explore this wicked complexity by first examining what is causing it and then discussing how conceptualizing the process of infrastructure change as satisficing numerous stakeholders and objectives, rather than simply optimizing a technical solution, is an increasingly important engineering framework, and hence professional skill.

Wicked Complexity

The process by which we design, build, operate, manage, rehabilitate, and decommission infrastructure is a wickedly complex problem. From an engineering perspective, infrastructure is often thought of and taught as a physical end product, an agglomeration of specialized technological and built artifacts that provide services. However, as technologies and the demand for services change more and more rapidly, thinking of infrastructure as a physical, cultural, and institutional process providing an ever-changing function becomes important.

This means that while an individual piece of physical infrastructure can be optimized for its function within an existing built system, infrastructure design and modernization necessarily becomes wickedly complex. Wicked complexity is the result of three competing forces that are, given the current approaches for designing and managing infrastructure, inimical to rapid and sustained change of infrastructure in a future marked by uncertainty. These forces are 1) wicked problems, 2) technical complexity and lock-in, and 3) social complexity, which work together to fragment our capacity to manage change (see Figure 1).

There is a dearth of work characterizing engineered infrastructure as wicked complex systems, but it is becoming more widely accepted that this is the case. The complexity of infrastructure has often been defined in terms of physical structure, or technical complexity.[16] The researchers Edward J. Oughton and Peter Tyler characterize infrastructure complexity by contrasting the properties of general and complex adaptive systems at the interface of supply and demand.[17] They posit that given emergent behavior and self-organization, instability and robustness, dynamics and evolution, adaptiveness, and agent diversity, infrastructure exhibit the key features of complex adaptive systems.

Figure 1. Wicked Complexity as a Product of Wicked Problems, Technical Complexity, and Social Complexity

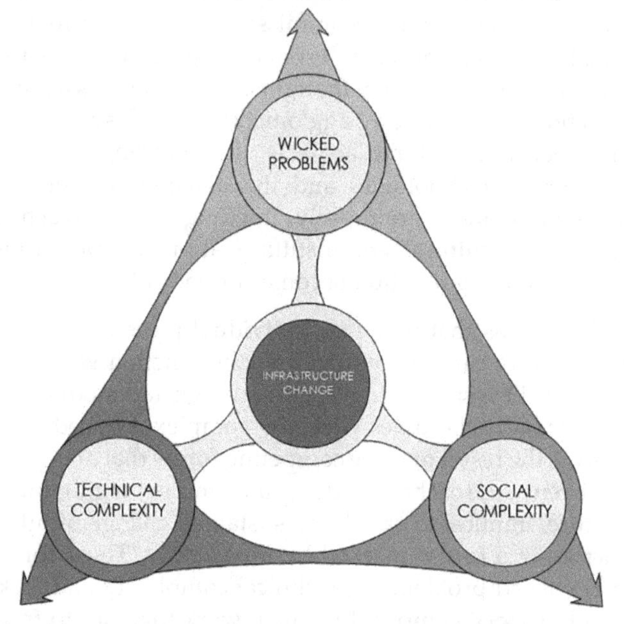

Adapted from J. Conklin, Dialogue Mapping: Building Shared Understanding of Wicked Problems *(Hoboken, NJ: Wiley, 2006).*

Complexity scientist Samuel Arbesman describes technical complexity as resulting from several dominant technological forces, including accretion (the accumulation and layering of technologies and assets over long periods of time), interaction (an often-indeterminate number of components interact in ways that we no longer fully comprehend), and edge cases (we sometimes design hardware and processes outside of standard design and operating rules).[18] The combination of these forces, and equally complex social, financial, cultural, technological, and management forces that affect infrastructure at every scale, results in a situation where the emergent behavior, particularly when perturbed, is no longer predictable. Defining infrastructure complexity in terms of physical structure is still appropriate for many purposes, of course, but rapidly becomes inadequate beyond the artifactual and short term, and especially when planning for new technologies.

There are many complexity definitions, often specific to the field and system or organization being analyzed. The Cynefin framework, created by management consultant Dave Snowden in 1999, classifies systems as simple (the domain of best practice), complicated (the domain of experts), complex (the domain of emergence), chaotic (the domain of rapid response), and disorder.[19] Knowing whether you're working in the complicated or complex domain when it comes to infrastructure is critical because each domain requires fundamentally different approaches.

The complex domain (the realm of unknown unknowns) is one where the inability to predict emergent behaviors means that we can only understand what happened after the fact. Emergence (the ability of individual components of a large system to work together to give rise to dramatic and diverse behavior) is often central to characterizing a system as complex.[20] Although emergence is front and center, complexity is determined from a number of characteristics, including the number of elements, interactions that are dynamic, rich, non-linear and short range, feedback loops, system openness, non-equilibrium behavior, histories, and (related to emergence) elements which are ignorant to the behavior of the whole system.[21]

The critical boundary for engineers is between complicated and complex. At this transition, optimization tools become obsolete as social, institutional, political, and economic forces come into play and together render insufficient quantitative analyses as the dominant approach for addressing the challenges. Organization scholar Jeffrey Conklin describes social complexity as a force of fragmentation that results from the number and diversity of players involved in a project (wickedness is another source of fragmentation in decision making).[22] The more parties involved and the more different the stakeholders are, the greater the diversity of perspective they bring, and the more socially complex a decision-making process becomes, thereby making collective intelligence a challenge and consensus virtually impossible to achieve.[23]

Figure 2. Cynefin Framework Applied to Infrastructure

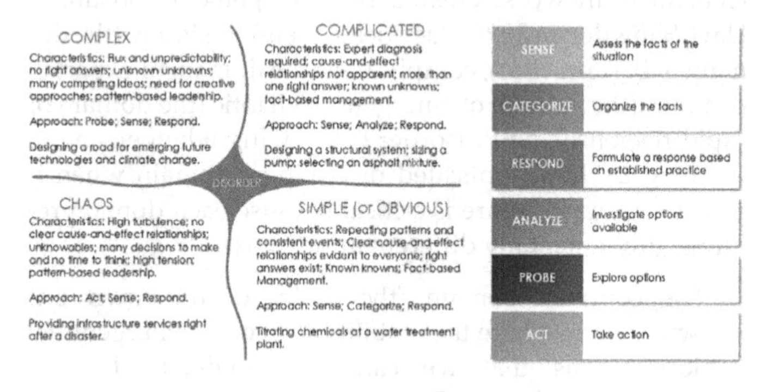

The somewhat ambiguous boundary between highly complicated technological artifacts or subsystems (for which optimization techniques are often appropriate) and wicked complexity (where satisficing techniques are necessary) occurs when operational requirements mean that infrastructure cannot simply be defined in terms of physical assets, but must be viewed as a process. Sometimes, after all, one is simply adding a component where the design is significantly constrained by existing function and assets—a tower to an existing urban mobile device network, or a new water feed to a subdivision.

But especially as human, natural, and built systems integrate in new ways at virtually all scales, the fundamental goal of infrastructure—to provide the resources and services needed to allow people the capabilities that build adaptive capacity and improve human well-being[24]—becomes more explicit. At this point, wicked complexity begins to dominate engineering decisions, from design to operation to maintenance to system evolution, and the lens of the engineer must broaden to consider not only the physical structures but also the institutional, cultural, financial, and other forces that affect infrastructure, both in the immediate environment and as the infrastructure process evolves over time.

The pioneering work of Horst W. J. Rittel and Melvin M. Webber provides the foundational criteria for wicked problems that for the most part endures today.[25] Instead of simply reiterating their 10 criteria, we'll instead describe them in the context of infrastructure:

1. Wicked problems have no definitive problem formulation. Providing services at reasonable costs, given financial constraints, competing stakeholder views, environmental and social implications of various options, and with agility and flexibility to adapt to changing conditions means that no clear formulation can be stated containing all of the information the decision maker needs to choose a single objective solution.

2. There is no stopping rule for infrastructure, which is an evolving process that must constantly be re-examined. A chosen infrastructure solution today is typically the result of a group of decision makers saying it's good enough. Had there been more time or resources, or different social or political constraints, a different solution may very well have been chosen.

3. Infrastructure solutions are not true-or-false, but better or worse, and each stakeholder may have a different evaluation of what these terms mean given the particular situation. It is usually not the case in these situations that a choice will be regarded as optimal by any party; rather, the best choice may be one that meets the minimal requirements of most stakeholders (in other words, the one that satisfices, rather than optimizes, from their perspective).

4. There is no immediate and no ultimate test of an infrastructure solution. The implementation of infrastructure will lead to direct and indirect consequences over an extended time period.

5. Given the sheer scale and reach of infrastructure, implementation is often a one-shot operation. There is no opportunity to learn by trial and error, and every attempt counts significantly. This does not mean, however, that engineers and engineering institutions such as universities and professional organizations should not explicitly embrace continual learning processes, because the effectiveness of a particular one-shot solution may well inform better practices elsewhere. For example, the Big Dig in Boston is a one-shot solution in that it is not significantly modifiable at this point, but the learning from what went well and didn't, and what performed socially and environmentally in the ways anticipated, can be applied to similar initiatives in, say, New York City or Toronto or Brasília.

6. There are no criteria by which we can say that all infrastructure solutions to a problem have been identified. As such, judgement dictates whether one should enlarge the set of solutions being considered.

7. Despite having implemented infrastructure solutions in similar forms for centuries, there is almost always a new set of distinguishing features for the next problem that makes it unique. This is especially true because, while purely technological elements of the system may be similar to predecessors, it is seldom the case that the environmental, social, political, and economic dimensions are the same. Rittel and Webber use a subway as an example. Using a subway system design that worked for one city as the template for another is likely to be disastrous, as differing commuter preferences and urban form may outweigh similarities in subway network structure.[26]

8. Change in infrastructure is often incremental, with the hope that each small step contributes to systematic improvement. However, this can be dysfunctional: tackling the need for change on such a small level can, by avoiding foundational improvement, even create more systemic inertia against fundamental, nonincremental change. Thus, for example, incremental improvement in the U.S. air traffic system has kept an obsolete infrastructure functioning, but the cost is continuing and growing inefficiency, and a more complex upgrade process when it is finally necessary.

9. The modes of reasoning for why we pick infrastructure solutions often far outweigh the choices available through scientific discourse. The problems that these solutions are attempting to address are often based on the most powerful decision makers' world views. When engineers and decision makers are not trained in satisficing techniques, or in recognizing the impact of their own worldviews, decisions are often costly and socially damaging (this was one of the critiques of the high modernists such as Robert Moses).

10. The goal of infrastructure is typically not to find some truth, but instead to improve some characteristic of the world. The problems that infrastructure managers must deal with don't have boundaries and their causes are often obfuscated.

The challenges associated with affecting change in infrastructures that are wicked and complex are exponentially greater and require fundamentally different approaches than what we've historically used. The technical and social forces of complexity combined with wicked infrastructure problems (including access to water, affordable housing, safe, efficient, and affordable mobility, and public health) result in a fragmentation of priorities driven by increasing numbers of stakeholders that render traditional approaches obsolete.

In the Anthropocene it appears that infrastructures are complex adaptive systems that engage wicked complexity on a routine basis. As such, agility and flexibility are needed because of unpredictable emergent properties, and wicked complexity adds the need for satisficing.

Infrastructure Satisficing

To understand how wicked complexity has emerged around infrastructure it's necessary to review the cultural history surrounding development. The modern era emerged at the end of the Renaissance and proceeds through the twentieth century.[27] It is largely associated with the Enlightenment: the development of scientific methods and an emphasis on empirical observation over reason and innate knowledge.[28] It is closely associated with the development of individualism, capitalism, expansion of world trade, urban development, and rapid technological progress.[29] So-called high modernism took root after World War II, emphasizing scientific and technological progress delivered by scientists, engineers, and bureaucrats attempting to master nature and largely ignoring the complex social, political, and economic underpinnings in development.[30] Development often focused on spatial ordering and disregarded history, geography, and social context.[31]

The high modernist period was the domain of the classically trained engineer and planner. Dominance of the political process and the top-down elitist approach characterized by Moses and Le Corbusier meant that integrated design optimized by the engineer was the model for successful infrastructure projects.[32] But high modernism failed, both because it didn't adequately serve an increasingly complex urban culture, and because it was too rigid and inflexible to meet the demands of increasingly vocal community and issue activists, such as environmentalists.[33]

In today's postmodernist environment, where many conflicting demands must be integrated in infrastructure projects, satisficing is the only viable methodology. This can be seen not just in infrastructure, but in the "adaptive management" methods that are becoming popular for managing "natural" infrastructure such as the Everglades.[34] The role of the infrastructure planner and designer has shifted from prescribing a professional solution to a given situation—a high modernist framing of the professional's role—to being the expert facilitator of the emergence of a design that satisfices enough stakeholders, along with the legacy systems and natural infrastructure components, to be both stable and viable.[35]

What the infrastructure manager needs to do today is very different than in the past. Tribalism is the new norm, where solutions are dictated not by technical performance measures but instead by "acceptable enough" to all parties.[36] While this approach has done a better job at bringing more voices to the table, it is inimical to flexibility and agility, the ability to change infrastructure quickly thereby creating adaptive capacity.

This is not to say that having community and multistakeholder perspectives involved in infrastructure decision making is necessarily a bad thing (this is crucial to ensuring that no party asserts their will over others). Instead, we must recognize that the fragmentation that comes with this tribalism means that there are competing narratives for what infrastructure should and should not do. A wicked problem is one where you don't understand the problem until you have a solution.

This social complexity creates situations where different stakeholders think that their problem is the only and right problem (e.g., the result of where they're from or the mission of their organization), and as such collective inquiry into the problem is prevented.[37] The stronger the network of stakeholders, the less agile and flexible the system.

The challenge is not to deny the involvement of stakeholders but to recognize that taken as a whole they are a significant design constraint that must be considered, and engineers must try to design for agility and flexibility in spite of it.

Managing Infrastructure in the Anthropocene

The processes of infrastructure design, delivery, management, and change will need to be as fast as the relevant changes in demand, technologies, demographics, and Earth systems in the Anthropocene. Optimizing, which is the current dominant paradigm, is wholly insufficient for the future. The fragmentation that has resulted from the wicked complexity that has emerged in the contemporary infrastructure process produces disincentives to embrace change and plan for uncertainty. In an increasingly divided and tribal society, even something as fundamental to infrastructure design and management as problem definition becomes ambiguous and contested.

Thus, rather than simply designing a solution to a given problem, the engineer will often find herself needing to satisfice both on problem definition and design solution. Wicked problems can fragment direction and mission since agreeing on the problem to develop a solution is challenging. Furthermore, social complexity fragments stakeholder unity through competing interests and identities. Given these challenges, we must seriously question whether optimizing as the dominant approach is sufficient in terms of significant infrastructure transformation for the Anthropocene. Even if we sought a different approach, the organizational competencies needed to address complexity in infrastructure institutions are not, for the most part, present.

A major challenge that infrastructure systems face in adapting is that their managing institutions are largely configured around bureaucracies designed to address complicated rather than complex systems. What's more, they typically encompass single disciplines (e.g., transportation, water, power, etc.).

Bureaucracies—characterized by hierarchical organizations, jurisdictional responsibilities, intentional and abstract rules, production and administration belonging to the office, appointed officials, and career employment—became commonplace during the industrial era.[38] Industrialization (in addition to representative governance and nationalism) added many new tasks to bureaucracies, transforming their role in the modern era. Public services, and their associated infrastructure, became necessary for democratizing governments during the colonial era, and bureaucracies became commonplace in the management of these services, many emerging during industrialization.[39]

The modern infrastructure system is mostly managed by bureaucratic structures that are designed for the industrial era, both in terms of the problems that we were facing and the intellectual needs to address those problems. The engineering education process largely reflects this. With the industrial era came a need to develop competencies around complicated problems.

Production became increasingly concentrated at factories (instead of agriculture) and the rapidly developing technologies and machinery of mass production needed workers who could understand and work with machines.[40] Hierarchies—the staple structure of bureaucracies—were used to organize tasks vertically within organizations establishing functional and social relationships.[41] Expertise was compartmentalized, access to information constrained, and resources earmarked for particular classes of workers.[42]

These hierarchies work against creativity: they emphasize efficiency of the status quo process. Ideas become stranded between levels of management within the hierarchy—each of which has competing priorities and doesn't necessarily have the authority to implement new ideas. And given the separation of problems, management often doesn't understand the value of the idea.

This organizational structure in many ways persists in current infrastructure management institutions, as well as in the educational structures that feed them. Compounding the traditional challenges is wicked complexity; not only are these bureaucracies expected to efficiently deliver public services but they must do so given the wicked problems, technical complexity, and social complexity that has become endemic.

Adaptive management starts at infrastructure institutions that embrace the competencies needed to deal with complexity, emphasize agility and flexibility of operations and physical hardware, and recognize that infrastructure as an ever-changing process. This kind of management will be necessary for the non-stationarity norms and rapidly changing needs of the Anthropocene.

Adaptive Management: Complicated to Complex Competencies

Infrastructure management in the Anthropocene will require new competencies that give systems the capacity to quickly adapt to deliver resources in non-stationarity and ever-more complex environments. The models of adaptive management and other successful strategies that have been developed for managing complexity should serve as a foundation for transforming the processes of infrastructure design, planning, operation, maintenance, and evolution.

Here we don't pretend to have a final vision of how this process is explicitly structured. On the contrary, we view the process of infrastructure as one that will need to evolve, and the forms that it takes at any given time in the future will need to reflect the complex environments and challenges of the time, most of which cannot be forecast, certainly not with any detail. However, the competencies associated with adaptive management strategies to address wicked complexity can serve as a guidebook.

There are lessons that can be learned for hard infrastructure systems from the practices developed by socio-ecological system (SES) practitioners. SES practitioners have been developing best practices for ecosystem resource management to coordinate resources in the face of complexity and uncertainty.[43] These adaptive management practices produce a "flexible system of resource management, tailored to specific places and situations, supported by, and working in conjunction with, various organizations at different scales."

The SES literature abounds with successful implementations of adaptive management for complex and rapidly changing social-ecological systems. Successes include the management of the Kristianstads Vattenrike region in Sweden, which balanced development, conservation, and ecological services to address water quality and biodiversity collapse; preservation of the Great Barrier Reef Marine Park in Australia, which was experiencing coral bleaching, overfishing, and eutrophication; and protection of fishing in the Southern Ocean surrounding Antarctica, which was experiencing fish stock collapse. The common thread in these examples was a looming crisis that triggered a few individuals to build trust and knowledge, connect networks, and develop a shared system vision.[44]

It is useful, for example, to compare recent adaptive management approaches to the Florida Everglades (challenged by phosphate mining, agricultural interests, and population growth and developers) with the experience of the Aral Sea (the Soviet Union diverted much of its inflow to growing cotton after World War II). Both regions are highly complex economically and culturally, but the use of adaptive management principles in the Everglades has prevented the economic, social, and ecological collapse that the Aral Sea region has suffered. A primary reason is that in the case of the Everglades, the wickedly complex nature of regional management was recognized, while in the case of the Aral Sea, the system was optimized for cotton production for export purposes.[45] It's critical that engineers and infrastructure managers develop competencies to be able to support and facilitate adaptive management practices.

First and foremost, infrastructure managers will need to be able to distinguish between complicated – even chaotic – but deterministic problems, and those characterized by wicked complexity. Distinctions will have to be made between systems where wicked complexity is relatively low, and systems where it is high. Engineers have in recent history largely operated in and been trained for the complicated domain – that of known unknowns – where management processes focus on unique skills associated with diagnosing and optimizing cause and effect relationships (e.g., what types of bridges could a city build across a canyon given geologic conditions, traffic requirements, and a budget). The complex domain is unique in that it's characterized by unpredictability and flux, which lead to emergence that cannot be predicted (unknown unknowns).[46] We won't know what will work and have to accept that the best that we'll be able to do is generate educated guesses and commit to a dynamic process where we constantly reassess what's happened and adjust course.

Examples of complex infrastructure challenges that face engineers are plenty. What is the right solution to a new tourism road through a Peruvian cloud forest for access to Incan ruins? The engineer must consider highly sensitive ecosystem impacts, preserving cultural heritage, design for climate change and possibly more extreme rainfall events, autonomous vehicles, among other factors. The likely answer is we don't know. Sure, we could build a typical asphalt road for modern automotive technology. But this road—given the unpredictability of the factors mentioned—may be insufficient for the conditions, technologies, and needs two decades from now. Instead, infrastructure processes that embrace complexity would focus on experimental management practices and a flexibility in understanding what does and does not work, and commit to trying something different later on once new information emerges.

Additionally, it's reasonable to expect infrastructure managers to have to operate in the domain of chaos more frequently in the future. Catastrophic events like climate change-driven natural hazards and terrorist attacks (particularly cyberattacks, given how interconnected physical infrastructure is becoming with information and communication technologies) are already becoming more common[47]—and revealing how little training our infrastructure managers have for working through these situations. In chaotic situations, infrastructure managers will need to be trained to help establish order, sense where stability is and is not present, and then help support transitions back to the complexity domain.[48] This will require training and practice that is often absent from traditional university curricula.

Managing complex infrastructure systems will require new approaches to training, education, and practice. The conventional infrastructure management organization is structured as a top-down hierarchy where expertise and resources are compartmentalized. Consider a hypothetical state department of transportation that has a leader at the top (often politically appointed), division directors that oversee various domains of the system (e.g., infrastructure delivery, planning, operations), and groups within those divisions that carry out the mission. In this structure, the sharing of knowledge and resources across groups to address interdisciplinary challenges is typically infeasible, and solutions to challenges are often prescribed with little opportunity for deviation. This structure emphasizes compartmentalization of knowledge and efficiency in solutions. More fundamentally, if a problem requires coordination beyond the transportation domain—say, implementing a tax structure that will encourage home office work rather than commuting to an office on bad air quality days—it is essentially impossible with the stove-piped structure of today's engineering management and education institutions.

There is an obvious need to develop organizational structures that 1) emphasize a diversity of ideas and perspectives, and offer frequent opportunities for exchanging them; 2) implement infrastructure systems as an experiment and dynamic reassessment exercise where patterns are allowed to emerge, are then studied, and new infrastructure is then implemented; and 3) emphasize the creation of new ideas over historical models. This organization would, for example, allow emerging technologies to be tested in practice, pit interdisciplinary teams against each other on the same problem, and provide workers discretionary time to pursue outside-the-box ideas as they relate to the agency's mission.[49]

Combining the emerging SES, management, and infrastructure concepts on wicked complexity, several core competencies emerge.

- *Shared Understanding*: Prior to managing wicked complex problems, there needs to be shared understanding of the problems and a willingness to address the problems collectively. Shared understanding is the process by which stakeholders are made aware of each other's goals and concerns; it is not consensus building. Shared understanding can lead to shared commitments on a project's directions, goals, emerging solutions, group decisions, and actions, and ultimately collective intelligence.[50] The processes by which infrastructure decisions are made, both internally and externally to the organization, should be radically altered toward collective intelligence. This is of course easier said than done, as a plethora of barriers exist that lock-in current practice and work against radical change. Nonetheless, as the forces increase that dictate how our future infrastructure can and cannot be in the Anthropocene, current practices that attempt to balance different constraints and perspectives without building collective intelligence appear to be increasingly insufficient.

- *Manage, Not Solve*: Infrastructure systems are too often treated as physical assets designed and operated to solve a problem, whether the facilitation of a volume of vehicle traffic or a minimum water pressure and volume. Yet in the face of increasing change, the solution will need to constantly evolve. As such, infrastructure managers will need to adopt a perspective that embraces change – that a system is temporary, the right fit for a short period of time, and likely unacceptable for the not-too-distant future. Management then focuses on the changing conditions that infrastructure needs to adapt to. This requires making decisions under deep uncertainty. In the past, infrastructure could be framed as artifactual; going forward, it must be reconceptualized as a process.[51]

- *Try, Learn, Adapt*: Complex environments are not conducive to command-and-control, top-down strategies; there's just too much distributed information, changing too rapidly, for a centralized approach to work. Instead, an approach that emphasizes experimentation as a process of learning to better enable adaptation capacities is needed. Here infrastructure organizations, the general public, and financiers need to allow for some level of failure. For example, a region could establish that a portion of transportation infrastructure funds be routinely used for testing for new technologies, such as alternative materials or traffic management, with the recognition that many of these technologies will not be implemented at scale but will provide valuable insights into what the agency needs to do to prepare for disruptive conditions. In addition, those responsible for infrastructure must learn to distribute management out to the network; decisions should be made at the most local level possible (a principle that in politics has become familiar as "subsidiarity," particularly in the European Union).

- *Complexity Mindset*: The training that many infrastructure managers receive focuses on managing complicated problems, often through the use of a predefined set of options or processes, and with explicit or implicit reference to optimization techniques. A complexity mindset accepts that complexity exists, that it needs to be managed differently, and that there are limitations on what a manager can control.[52] A complexity mindset focuses on what can be over what is, and relies on satisficing, not optimizing, mental models and methods. Planning is different in each case. With complicated systems, planning is done to determine paths which are then followed. With complexity, it is the process of constantly planning that informs action; the plans themselves are seldom, if ever, implemented. As General (later President) Eisenhower observed, "In preparing for battle I have always found that plans are useless, but planning is indispensable."[53] It isn't the plan that's important; it is the act of planning—thinking about the possibilities, the interconnectedness of problems, and emerging properties of the system—that's not just necessary but a critical part of adaptive management.

Every indication is that the competing forces that now define how infrastructure is implemented, and that are driving increasing complexity today, will continue if not accelerate. Unpredictable and accelerating evolution is occurring across the entire technological frontier. Social and political fragmentation and complexity is growing, and the stability of a world order which has prevailed for over three centuries is failing.[54] Environmental pressures and associated perturbations in human demographics and migration patterns, as well as unpredictable shifts in natural cycles and systems, are set to accelerate. Urbanization is accelerating, especially in developing countries.

There is thus little question that demands on technologists, managers, and engineers working on infrastructure systems are also going to grow. New approaches are needed to create coherence. For example, if water demand is increasing, then a treatment plant will need to be created based on the forecasted demand; funding may need to be secured from a new tax measure; political and social factors will need to managed; and water rights may need to be secured. Acceptance by local communities, activist organizations (which often have different if not mutually exclusive agendas), and regional water users and managers will be required. Designs that are flexible enough to adapt to unpredictable changes in supply and demand because of climate change must be developed, and metrics for tracking performance of the infrastructure—and the systems context of the infrastructure—must be developed and institutionalized. The complicated task of designing the water treatment plant is possibly the simplest exercise in the whole process, and is perhaps the only one for which the infrastructure manager has been trained.

Conclusion

In the Anthropocene, infrastructure is an integration of co-evolving human, built, and natural systems, and inevitably is characterized by wicked complexity. As such, engineers and infrastructure managers will need to shift their mindset away from optimizing and toward satisficing. They will need to shift their thinking about infrastructure from the delivery of physical assets that meet a defined and relatively stable need to a process by which physical systems are designed, built, and operated in response to shifting priorities, new technologies and social practices, and irreducible uncertainty.

Notes

[1] H. Simon, *Administrative Behavior: A Study of Decision-making Processes in Administrative Organization* (Basingstoke, UK: Macmillan Company, 1947); H. Simon, *Models of Man: Social and Rational* (Berkeley, CA: Wiley, 1957); R. Brown, (2004) "Consideration of the origin of Herbert Simon's theory of 'satisficing' (1933-1947)," *Management Decision* 42, no. 10 (2004): 1240–1256; "Herbert Simon," *The Economist* (March 20, 2009).

[2] H. Simon, *Models of Man: Social and Rational* (Berkeley, CA: Wiley, 1957); R. Brown, "Consideration of the origin of Herbert Simon's theory of 'satisficing' (1933-1947)," *Management Decision* 42, no. 10 (2004): 1240–1256.

[3] H. Simon, *Models of Man: Social and Rational* (Berkeley, CA: Wiley, 1957).

[4] A. Carse, "Keyword: infrastructure: How a humble French engineering term shaped the modern world," in *Infrastructures and Social Complexity*, eds. P. Harvey, C. Bruun Jensen, and A. Morita (Oxfordshire, UK: Routledge, 2017), 45–57; S. S. Clark, T. P. Seager, and M. V. Chester, "A capabilities approach to the prioritization of critical infrastructure," *Environment Systems and Decisions* 38 (2018): 339-352.

[5] M. Chester and B. Allenby, "Towards Adaptive Infrastructure: Flexibility and Agility in a Non-Stationarity Age," *Sustainable and Resilient Infrastructure* 3, no. 1 (2018): 1–19.

[6] Ibid.

[7] Ibid.; M. V. Chester, S. A. Markolf, and B. Allenby, "Infrastructure and the Environment in the Anthropocene," *Journal of Industrial Ecology* 23, no. 5 (2019): 1006–1015.

[8] M. Chester and B. Allenby, "Towards Adaptive Infrastructure: Flexibility and Agility in a Non-Stationarity Age," *Sustainable and Resilient Infrastructure* 3, no. 1 (2018): 1–19.

[9] P. Bogucki, "How Wealth Happened in Neolithic Central Europe," *Journal of World Prehistory* 24 (2011): 107–115.

[10] J. Gies, *Cathedral, Forge, and Waterwheel* (New York, NY: HarperCollins, 1995).

[11] N. Rosenberg and L. E. Birdzell, *How the West Grew Rich: The Economic Transformation Of The Industrial World* (New York, NY: Basic Books, 1986); R. Kurzweil, *The Singularity is Near* (London, UK: Gerald Duckworth & Co., 2010); G. E. Marchant, B. R. Allenby, and J. R. Herkert, *The Growing Gap Between Emerging Technologies and Legal-Ethical Oversight: The Pacing Problem* (New York, NY: Springer, 2011); B. Allenby, *The Theory and Practice of Sustainable Engineering* (Upper Saddle River, NJ: Pearson Prentice Hall, 2012).

[12] P. Hall, *Cities in Civilization: Culture, Innovation, and Urban Order* (London, UK: Weidenfeld & Nicolson, 1998).

[13] M. Berman, *All That Is Solid Melts into Air: The Experience of Modernity* (New York, NY: Simon and Schuster, 1982).

[14] J. Jacobs, *The Death and Life of Great American Cities* (New York, NY: Vintage Books, 1961); C. Jencks, *The Language of Post-modern Architecture* (New York, NY: Academy Editions, 1977).

[15] P. Hall, *Cities in Civilization: Culture, Innovation, and Urban Order* (London, UK: Weidenfeld & Nicolson, 1998); M. Gandy, *Concrete and Clay: Reworking Nature in New York City* (Cambridge, MA: MIT Press, 2003).

[16] T. Brown, W. Beyeler, and D. Barton, "Assessing infrastructure interdependencies: the challenge of risk analysis for complex adaptive systems," *International Journal of Critical Infrastructures* 1, no. 1 (2004): 108–117; T. D. O'Rourke, "Critical Infrastructure Interdependencies, and Resilience," *The Bridge* 37, no. 1 (2007): 22–29.

[17] E. Oughton, W. Usher, J. Hall, and P. Tyler, "Infrastructure as a Complex Adaptive System," *Complexity* 1 (2018).

[18] S. Arbesman, *Overcomplicated: Technology at the Limits of Comprehension* (New York, NY: Penguin Publishing, 2016).

[19] D. J. Snowden and M. E. Boone, "A Leader's Framework for Decision Making," *Harvard Business Review* (Nov. 2007).

[20] P. Cilliers, *Complexity and Postmodernism: Understanding Complex Systems* (London, UK: Taylor & Francis, 2002); A. Martin and K. Helmerson, "Emergence: the remarkable simplicity of complexity," *The Conversation* (Sept. 30, 2014).

[21] P. Cilliers, *Complexity and Postmodernism: Understanding Complex Systems* (London, UK: Taylor & Francis, 2002).

[22] J. Conklin, *Dialogue Mapping: Building Shared Understanding of Wicked Problems* (Hoboken, NJ: Wiley, 2006).

[23] Ibid.

[24] S. S. Clark, T. P. Seager, and M. V. Chester, "A capabilities approach to the prioritization of critical infrastructure," *Environment Systems and Decisions* 38 (2018): 339–352.

[25] H. Rittel and M. Webber, "Dilemmas in a general theory of planning," *Policy Sciences* 4, no. 2 (1973): 155–169.

[26] Ibid.

[27] P. J. Hugill, *World Trade Since 1431: Geography, Technology, and Capitalism* (Baltimore, MD: Johns Hopkins University Press, 1995).

[28] D. S. Landes, *The Wealth and Poverty of Nations: Why Some are So Rich and Some So Poor* (New York, NY: W. W. Norton & Company, 1998).

[29] R. Findlay and K. H. O'Rourke, *Power and Plenty: Trade, War, and the World Economy in the Second Millennium* (Princeton, NJ: Princeton University Press, 2007); I. Morris, *Why The West Rules – For Now: The Patterns of History and What They Reveal about the Future* (New York, NY: Farrar, Straus, and Giroux, 2010).

[30] M. Berman, *All That Is Solid Melts into Air: The Experience of Modernity* (New York, NY: Simon and Schuster, 1982).

[31] J. Jacobs, *The Death and Life of Great American Cities* (New York, NY: Vintage Books, 1961).

[32] C. Jencks, *The Language of Post-modern Architecture* (New York, NY: Academy Editions, 1977); M. Berman, *All That Is Solid Melts into Air: The Experience of Modernity* (New York, NY: Simon

and Schuster, 1982); L. Corbusier, *The City of To-morrow and Its Planning* (New York, NY: Dover, 1987).

[33] C. Jencks, *The Language of Post-modern Architecture* (New York, NY: Academy Editions, 1977); P. Hall, *Cities in Civilization: Culture, Innovation, and Urban Order* (London, UK: Weidenfeld & Nicolson, 1998).

[34] F. Berkes, C. Folke, and J. Colding, *Linking Social and Ecological Systems: Management Practices and Social Mechanisms for Building Resilience* (Cambridge, UK: Cambridge University Press, 1998); B. Allenby, *The Theory and Practice of Sustainable Engineering* (Upper Saddle River, NJ: Pearson Prentice Hall, 2012); National Research Council, *Progress Toward Restoring the Everglades: The Seventh Biennial Review – 2018* (Washington, DC: National Academies Press, 2018).

[35] B. Allenby, *The Theory and Practice of Sustainable Engineering* (Upper Saddle River, NJ: Pearson Prentice Hall, 2012).

[36] A. Chua, *Political Tribes: Group Instinct and the Fate of Nations* (New York, NY: Penguin Random House, 2018); S. Hawkins, D. Yudkin, M, Juan-Torres, and T. Dixon, *Hidden Tribes: A Study of America's Polarized Landscape* (New York, NY: More in Common, 2018).

[37] J. Conklin, *Dialogue Mapping: Building Shared Understanding of Wicked Problems* (Hoboken, NJ: Wiley, 2006); B. Allenby, *The Theory and Practice of Sustainable Engineering* (Upper Saddle River, NJ: Pearson Prentice Hall, 2012).

[38] H. Constas, "Weber's Two Concepts of Bureaucracy," *American Journal of Sociology* 63, no. 4 (1958): 400–409; M. Weber, *The Theory of Social and Economic Organization* (New York, NY: Free Press, 2009); T. Diefenbach and R. T. By, "Bureaucracy and Hierarchy—What Else!?" in *Reinventing Hierarchy and Bureaucracy–from the Bureau to Network Organizations* (Bingley, UK: Emerald Group Publishing, 2012), 1–27.

[39] F. W. Riggs, "Modernity and Bureaucracy," *Public Administration Review* 57, no. 4 (1997): 347.

[40] R. Nason, *It's Not Complicated: The Art and Science of Complexity for Business* (Toronto, Canada: University of Toronto Press, 2017).

[41] M. Weber, T. Parsons, and R. H. Tawney, *The Protestant Ethic and the Spirit of Capitalism* (New South Wales, Australia: G. Allen & Unwin, 1930).

[42] T. Diefenbach and R. T. By, "Bureaucracy and Hierarchy—What Else!?" in *Reinventing Hierarchy and Bureaucracy–from the Bureau to Network Organizations* (Bingley, UK: Emerald Group Publishing, 2012), 1–27.

[43] D. R. Armitage, R. Plummer, F. Berkes, R. I. Arthur, A. T. Charles, I. J. Davidson-Hunt, A. P. Diduck, N. C. Doubleday, D. S. Johnson, M. Marschke, P. McConney, E. W. Pinkerton, and E. K. Wollenberg, "Adaptive co-management for social-ecological complexity," *Frontiers in Ecology and the Environment* 7, no. 2 (2009): 95–102; B. C. Chaffin, H. Gosnell, and B. A. Cosens, "A decade of adaptive governance scholarship: synthesis and future directions," *Ecology and Society* 19, no. 3 (2014): 56.

[44] L. Schultz, C. Folke, H. Österblom, and P. Olsson, "Adaptive governance, ecosystem management, and natural capital," *Proceedings of the National Academy of Sciences* 112, no. 24 (2015): 7369–7374.

[45] B. Allenby, *The Theory and Practice of Sustainable Engineering* (Upper Saddle River, NJ: Pearson Prentice Hall, 2012); National Research Council, *Progress Toward Restoring the Everglades: The Seventh Biennial Review – 2018* (Washington, DC: National Academies Press, 2018).

[46] D. J. Snowden and M. E. Boone, "A Leader's Framework for Decision Making," *Harvard Business Review* (Nov. 2007).

[47] P. W. Singer and A. Friedman, *Cybersecurity: What Everyone Needs to Know* (New York, NY: Oxford University Press USA, 2014); USGCRP, *Climate Science Special Report: Fourth National Climate Assessment* (Washington, DC: U.S. Global Climate Change Research Program, 2017); M. Roser, M. Nagdy, and H. Ritchie, "Terrorism," *Our World in Data* (July 2013).

[48] D. J. Snowden and M. E. Boone, "A Leader's Framework for Decision Making," *Harvard Business Review* (Nov. 2007).

[49] Ibid.

[50] J. Conklin, *Dialogue Mapping: Building Shared Understanding of Wicked Problems* (Hoboken, NJ: Wiley, 2006).

[51] B. Allenby, *The Theory and Practice of Sustainable Engineering* (Upper Saddle River, NJ: Pearson Prentice Hall, 2012).

[52] R. Nason, *It's Not Complicated: The Art and Science of Complexity for Business* (Toronto, Canada: University of Toronto Press, 2017).

[53] R. Nixon, *Six Crises* (New York, NY: Simon & Schuster, 2013).

[54] F. Fukuyama, *Political Order and Political Decay: From the Industrial Revolution to the Globalization of Democracy* (New York, NY: Farrar, Straus, and Giroux, 2014); H. Kissinger, *World Order* (New York, NY: Penguin Publishing, 2014).

ACKNOWLEDGEMENTS

This book was supported in part by U.S. National Science Foundation grants, including the Urban Resilience to Extremes Sustainability Research Network (Award No. 1444750) and Growing Resilience Convergence Research (Award No. 1934933).

ACKNOWLEDGEMENT

ABOUT THE AUTHORS

Mikhail Chester

Mike Chester is the director of the Metis Center for Infrastructure and Sustainable Engineering at Arizona State University, where he maintains a research program focused on preparing infrastructure and their institutions for the challenges of the coming century. His work spans climate adaptation, disruptive technologies, innovative financing, transitions to agility and flexibility, and modernization of infrastructure management. He is broadly interested in the changes needed in infrastructure governance, design, and education for the Anthropocene, an era marked by acceleration and uncertainty. He is co-lead of the Urban Resilience to Extremes Sustainability Research network, composed of 19 institutions and 250 researchers across the Americas, focused on developing innovative infrastructure solutions for extreme events. He was awarded the American Society of Civil Engineer's early career researcher Huber prize in 2017. He recently published *Urban Infrastructure: Reflections for 2100* (2020).

Braden Allenby

Brad Allenby is President's Professor of Engineering and Lincoln Professor of Engineering and Ethics at Arizona State University. He moved to ASU from his previous position as the environment, health, and safety vice president for AT&T in 2004. Allenby received his BA from Yale University, his JD and MA (economics) from the University of Virginia, and his MS and PhD in environmental sciences from Rutgers University. He has served as president of the International Society for Industrial Ecology; chair of the AAAS Committee on Science, Engineering, and Public Policy; and chair of the IEEE Presidential Sustainability Initiative. He is also a AAAS Fellow and a Fellow of the Royal Society for the Arts, Manufactures & Commerce. Recent books include *The Techno-Human Condition* (with Daniel Sarewitz, 2011), *The Theory and Practice of Sustainable Engineering* (2011), *Future Conflict & Emerging Technologies* (2016), and *The Applied Ethics of Emerging Military and Security Technologies* (2016).

CPSIA information can be obtained
at www.ICGtesting.com
Printed in the USA
FSHW012010050122
87442FS